Systems Cost Engineering

Systems Cost Engineering

Program Affordability Management and Cost Control

DALE SHERMON

LONDON AND NEW YORK

First published 2009 by Gower Publishing

2 Park Square, Milton Park, Abingdon, Oxon OX14 4RN
711 Third Avenue, New York, NY 10017, USA

Routledge is an imprint of the Taylor Francis Group, an informa business

First issued in paperback 2016

Copyright © 2009 Dale Shermon

Dale Shermon has asserted his moral right under the Copyright, Designs and Patents Act, 1988, to be identified as the author of this work.

All rights reserved. No part of this book may be reprinted or reproduced or utilised in any form or by any electronic, mechanical, or other means, now known or hereafter invented, including photocopying and recording, or in any information storage or retrieval system, without permission in writing from the publishers.

Notice:
Product or corporate names may be trademarks or registered trademarks, and are used only for identification and explanation without intent to infringe.

British Library Cataloguing in Publication Data
Shermon, Dale.
　Systems cost engineering : program affordability management and cost control.
　1. Systems engineering--Cost control--Mathematical models.
　2. Parameter estimation. 3. Value analysis (Cost control)--Mathematical models.
　I. Title
　658.1'552'015118-dc22

　ISBN 978-0-566-08861-2 (hbk)
　ISBN 978-1-138-25386-5 (pbk)

Library of Congress Cataloging-in-Publication Data
Shermon, Dale.
　Systems cost engineering : program affordability, management, and cost control / by Dale Shermon.
　　p. cm.
　Includes bibliographical references and index.
　ISBN 978-0-566-08861-2
　1. Systems engineering--Costs. 2. Systems engineering--Management. I. Title.
　TA168.S44153 2009
　658.15'52--dc22

2009005471

Contents

List of Figures   *vii*
List of Tables   *xi*
Foreword   *xiii*
Acknowledgements   *xv*
Glossary   *xvii*

Chapter 1	Introduction	1
Chapter 2	How to … Appreciate Parametrics	5
Chapter 3	How to … Estimate Using Parametrics	23
Chapter 4	How to … Prepare Bids Faster with Fewer Resources	31
Chapter 5	How to … Prepare a Focused Business Plan	41
Chapter 6	How to … Validate Quotations from Suppliers	49
Chapter 7	How to … Manage a Program Effectively	57
Chapter 8	How to … Achieve Accuracy in Cost Engineering	71
Chapter 9	How to … Accomplish Quality Assurance	85
Chapter 10	How to … Estimate Through Life	105
Chapter 11	How to … Estimate Technology Maturity	111
Chapter 12	How to … Assess Software	127
Chapter 13	How to … Analyse Risk and Uncertainty	141

Chapter 14	How to ... Influence Project Strategy	159
Chapter 15	How to ... Consider Technology Insertion	181
Chapter 16	How to ... Develop Cost-effective Alternatives	195
Chapter 17	How to ... Tackle the System of Systems Challenge	207
Chapter 18	How to ... Create Home-grown Parametric Models	221
Chapter 19	How to ... Successfully Conduct Life-Cycle Costing	241
Chapter 20	How to ... Accomplish Knowledge Retention	253
Chapter 21	How to ... Present the Results	265
Chapter 22	How to ... Adopt Parametrics	273
Chapter 23	The History of Parametrics	291
Index		*301*

List of Figures

1.1	Program Affordability Management (PAM)	2
1.2	Example of documentation structure	3
2.1	The advantage of parametrics	6
2.2	Parametric model example	6
2.3	Enterprise applications	7
2.4	Example parametric cost model – Product Breakdown Structure (PBS), Cost Drivers and outputs	9
2.5	Normalized cost as a function of weight and complexity	12
2.6	Manufacturing Complexity attributes	14
2.7	Simple representation of a parametric hardware model	16
2.8	Determining an average Manufacturing Complexity	17
2.9	Estimating with the average Manufacturing Complexity	18
2.10	SUM formula in Excel	19
2.11	TrueAnalyst from PRICE Systems open architecture with visibility of CERs	20
3.1	First step – estimating	24
3.2	Second step – first level parameters	25
3.3	Third step – second level calculator inputs	26
3.4	Run the model and outputs	27
3.5	Elements in a parametric model	29
4.1	Rapid estimating with parametrics	32
4.2	Iterative proposal preparation process	33
4.3	Modelling schedule penalties	35
4.4	Comparison of Suppliers A and B	37
4.5	Example parametric questionnaire	38
5.1	Productivity analysis compared to average industry	43
5.2	Productivity tracking	45
5.3	Product cut-away	46
5.4	Missile PBS	47
6.1	Establishing a complexity for a family of technologies	50

6.2	Looking for explanations in the design or requirements	50
6.3	Example supply chain	53
6.4	Closing the cost engineering and procurement loop	54
7.1	Definition of a Control Account	58
7.2	The time, cost, performance triangle	59
7.3	Integrated EVM and estimating system	62
7.4	Predictive EVM integration	64
7.5	Initial baseline estimate	65
7.6	Progress at seven months	66
7.7	Predictive EVM prediction	67
8.1	Calibration	74
8.2	Results of calibration	75
8.3	Identifying technology families	76
8.4	Fifty calibrated structural items	78
8.5	Organizational calibration	81
9.1	Multiple estimating methods	86
9.2	Compounding errors in estimating	90
9.3	PBS detail and accuracy	91
9.4	Fighter aircraft Historical Trend Analysis	94
9.5	Client/Server Administration screen by PRICE Systems	98
9.6	Client/Server security	99
9.7	Computer Aided Design (CAD) and parametric estimating integration	102
10.1	Through Life Estimating methodologies	107
10.2	Integration of different estimating methods	108
10.3	The desired outcome of Through Life Estimating	109
11.1	The Technology Readiness Level (TRL) scale	112
11.2	Engineering Complexity Calculator	115
11.3	TRL versus Engineering Complexity	116
11.4	Technology Maturity model	119
11.5	Technology Historical Trend Analysis (HTA)	120
11.6	TRL phases	120
11.7	Progressive refinement and merger of technology and research	121
11.8	TRL Product Breakdown Structure (PBS)	122
11.9	MS-Excel link calculating the number and type of prototypes	123
11.10	Sample results for schedule demonstration	126
12.1	Software size	128
12.2	Sample inheritance tree	135
12.3	Use Case calculator	137

12.4	Simple retail system	138
13.1	Cause and effect diagram	143
13.2	Consequences of risk	144
13.3	The risk process	145
13.4	Probability Impact Grid (PIG)	148
13.5	The four steps of risk methodology	151
13.6	Example risk analysis inputs	152
13.7	Risk analysis correlation across the PBS	153
13.8	Risk output	154
13.9	Defining optimistic and pessimistic values	155
13.10	Results of analysis	156
14.1	Multinational projects	165
14.2	Technology maturity related to acquisition	166
14.3	Schedule cost penalties	168
14.4	Relative relationship to program phases	169
14.5	Continuous production or split production	170
14.6	Batch quantities	171
14.7	Benefits from cost improvement curves	172
14.8	Benefits of the quantity allocated to the appropriate project	172
14.9	Life Cycle Cost capability of parametric models	173
14.10	Possible hardware life-cycle maintenance concept	175
14.11	Extrapolation of military aircraft radio technology	176
14.12	Top level technology insertion program	177
14.13	Time-phased expenditure	178
15.1	Acquisition and procurement definitions	182
15.2	UAV options	185
15.3	Option 1 – hardware and software PBS	186
15.4	Option 0 – degrees of technology insertion	187
15.5	Option 1 – degrees of technology insertion	188
15.6	Possible technical architecture of receptive and non-receptive systems	189
15.7	Historical Trend Analysis	190
15.8	Comparison of Options 0 and 1 cost profiles	192
16.1	The top level process	197
16.2	Trade studies process	200
16.3	Performance driven gas turbine complexity calculator	201
16.4	Trade study considerations	204
16.5	Enterprise integration	205
17.1	System of equipments	208
17.2	System of Systems project structure	209

17.3	Lead System Integrator (LSI) and primes	214
18.1	Through Life Modelling capability	224
18.2	Parametric cost modelling	225
18.3	Catalogue cost research	226
18.4	Example of a cost model database and an indication of its coverage	227
18.5	Multi-colinearity	230
18.6	Weight equation determines T_1	231
18.7	Residual (normalized history less prediction) versus In-Service Date	232
18.8	Considering technology impact on systems of the same weight	233
18.9	Combined effect of the three hypotheses	234
18.10	Demand and schedule	235
18.11	Non-recurring development costs	236
18.12	TrueAnalyst – equation visibility	237
18.13	Model testing	238
18.14	Non-cost performance validation	240
19.1	The example questionnaire	249
19.2	The questionnaire	250
19.3	Questionnaire process	251
20.1	Security	257
20.2	Aircraft carriers retrieved	261
20.3	Analysis of aircraft carriers	262
21.1	Derivation of the estimate	267
21.2	The data centre problem	269
21.3	Graphical presentation of cost-versus-benefit analysis	270
21.4	Constant cost or constant performance	271
22.1	User requirements database	275
22.2	RACI matrix	286
22.3	Example client/server configuration	288
23.1	Frank Freiman – founder of parametrics	294
23.2	The Freiman Curve	295
23.3	Commercial software parametric models	297
23.4	Commercial hardware parametric models	298

List of Tables

8.1	Example of calibrated accuracy	72
8.2	Average complexity values for families of technologies	76
8.3	Reference range per family of technology	77
8.4	Comparison of calibrated and calculated complexities	79
8.5	Analysis of accuracy relative to accuracy definitions	80
11.1	Number of prototypes	124
11.2	New design	125
11.3	Engineering Complexity	125
13.1	Risk identification techniques	146
13.2	Top risks	157
19.1	Advantages and disadvantages of data gathering methods	245
21.1	Data centre problem	270
22.1	Dealing with reactions to change	274

Foreword

By Tony DeMarco
President, PRICE Systems

I am an entrepreneur, but my academic training was in mathematics and computer science, not business or accounting. One day I sat down with a pencil and a blue-lined paper pad (I could not find my graph paper) to craft a formula to determine the amount of near liquid cash we had to meet our operating needs. I laboured over income statements and balance sheets for a day before I found an equation that I knew would work. Proud of my accomplishment, I showed this financial revelation to our accountant. He promptly told me that it was the formula for Working Capital found in every beginners accounting text book. I learnt; there were better ways to have spent my day.

Program Affordability Management is the set of coordinated activities that determine whether or not an organization will be able to bear the cost of a program over the course of its life. It takes leadership, discipline and people armed with effective methods and tools to practise Program Affordability Management successfully. Great tools alone will not keep programs affordable. Tools must be applied as part of a credible process if estimates and analyses are to be accepted. We want people to be successful with the tools and solutions they use, so this book is a collection of methods with proven success.

Consider the needs of your organization and challenge people 'why are we not performing these activities?' Don't reinvent the wheel or accounting equations, learn from others. Familiarize yourself with this book's contents and keep it by your side. Your days will be more productive.

Acknowledgements

This book was inspired during work on more than 25 years of conference and symposium papers written by customers and consultants on the applications of parametric cost models.

The inspiration to compile a book came from Dale Shermon, but the contents have been gathered from the team of consultants at PRICE Systems with a collective Cost Engineering practical knowledge in excess of 300 years.

Therefore, acknowledgement is extended to the following team who assisted in writing this book:

Didier Barrault – France

Bill Mathis – USA

Anthony DeMarco – USA

Jeff Murphy – USA

Ron Dias – USA

Kevin McKeel – USA

Fabian Eilingsfeld – Germany

Arlene Minkiewicz – USA

Bruce Fad – USA

Jim Otte – USA

Pascal Gendrot – France

Peter Pizzutillo – USA

Bob Green – USA

Shamraz Razzaq – UK

Grahame Jones – UK

Larry Reagan – USA

Zach Jasnoff – USA

David Seaver – USA

Robert Kennedy – USA

Dale Shermon – UK

John Long – USA   Pete Stanley – USA

Emmanuel Mary – France   Georges Teologlou – France

'The team is greater than the sum of its parts.'

Glossary

Calibration	The process of tuning a commercial parametric model to an individual organization by producing the productivity metrics of historical projects.
Commercial Off the Shelf (COTS)	Referring to items which are purchased or built to licence or to the design of a customer. They can be either software or hardware items.
Manufacturing Complexity	An empirical factor comprising the technology implicit in a product and the productivity of its manufacturer. Most easily perceived as a normalized cost density in a hardware parametric model.
Organizational Breakdown Structure (OBS)	This is a formal arrangement of resources (labour and non-labour) which will need to be consumed or used to ensure successful completion of the project.
Organizational Productivity	The calibration factor for a software parametric model which represented the efficiency or productivity of an organization in software projects.

Product Breakdown Structure (PBS)	This is a formal arrangement of technologies or software which will need to be acquired or built to ensure successful completion of the project.
Program	An alternative description of a project taking into account all its facets including the budget and schedule.
Programme	A software code used to make computers perform a useful function.
Work Breakdown Structure (WBS)	This is a formal arrangement of activities or tasks which will need to be conducted to ensure successful completion of the project.

1

Introduction

Cost Engineering requires the fusion of three elements: processes, cost models and skilled people. When these three elements are combined efficiently, a capability is achieved that will profoundly influence the projects that an organization embarks upon. When these elements are realized in the organization, then cost estimating naturally leads to project control, which enables the development of corporate knowledge and the re-use of what has been learned in the cost estimates of the future.

Program Affordability Management (PAM) (see Figure 1.1) is a seamless union of these elements that results in what we call True Program Success. How do we know when we have achieved True Program Success? When we can confidently say, no program will ever:

- be conceived without a credible analysis of alternatives;

- be initiated with insufficient funding because of inaccurate initial estimates and inaccurate quantification of the risks;

- be deterred from its mission because of lack of credible cost analysis within the program's management;

- be deterred from its mission because of lack of integration between Earned Value Management and Cost Estimating and Analysis;

- be deterred from its mission because knowledge of cost and productivity metrics is not being shared among program teams and with other programs;

- be deterred from its mission because of surprise cost overruns and schedule delays.

Figure 1.1 Program Affordability Management (PAM)

Program Affordability is achieved through three elements of Cost Engineering:

1. *TrueMethods* – which are best practices in Advanced Planning, Bid and Proposal Development, Supplier Assessment and Selection, and Project Cost Control. These are the processes which are applications of a cost model taught by experts and applied by the relevant staff.

2. *TruePlanning* – which is a comprehensive cost analysis and knowledge management tool. It is a cost model framework which can be applied in numerous applications with the right training and experience.

3. *Parametric consultants* – who are experts in building budgets, evaluating performance, identifying risks, analysing tradeoffs and continuously monitoring program value. They are able to customize standard processes and training, to suit the environment, with models that relate directly to the organization. Alternatively, they can mentor the relevant staff in the same tasks.

Purpose of this Book

The purpose of this book is to describe how to achieve applications of cost and schedule models within an organization. The applications described in the following chapters have evolved over many years, until they have become elements of the philosophy of Program Affordability Management (PAM).

Figure 1.2 shows the structure of the documentation which parametric cost model vendors provide. The Graphical User Interface (GUI) Guide details the software and its use in terms of file storage, retrieval and screen manipulation. The model references are used to describe the characteristics of those models: their inputs, outputs and functional relationships – what makes them tick.

This book is an overarching guide to the application of the individual cost models. Independently of the parametric model being used, its intention is to provide solutions to the cost estimating problems posed in businesses today; enabling a return on the investment for a parametric model.

Each of the following chapters tackles a different application of the parametric methodology of cost and schedule estimating. As such, it is independent of the specific cost model used.

Frequently the terms Cost Estimator, Cost Analyst, Parametrician, Cost Engineer, Cost Forecaster and others are used interchangeably. In some organizations these terms have very specific and defined meanings. In this book

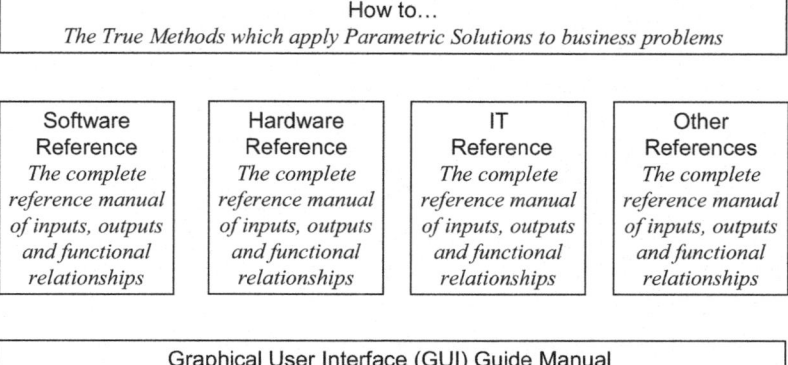

Figure 1.2 Example of documentation structure

the term Cost Engineer will be consistently used to describe a trained person able to operate, interpret and calculate a parametrically generated estimate. This is not intended to exclude staff who do not have the title Cost Engineer, but is simply designed to make the book more consistent.

2

How to ... Appreciate Parametrics

Why do so many large organizations around the world use parametrics? The reason is to provide speed in estimating the cost and schedule duration when little information exists (see Figure 2.1) regarding a project, proposed program, competitors product or suppliers equipment. During the early stages it is possible to influence the design decisions that are being made and thus reduce the cost of the technology being designed. Parametrics does not necessarily provide more accuracy; although it can contribute when combined with other methods, as we will see in later chapters.

Parametrics is a similar method of estimating to those used by any Cost Engineer who has constructed a spreadsheet model. When constructing a cost model the first step is to gather an international database of historical past projects. From this database, the Cost Drivers – those design and performance characteristics that influence the out-turn project costs, schedule and performance – can be observed. This enables the establishing of equations or algorithms which relate the Cost Drivers to the outputs. These are usually referred to as Cost Estimating Relationships or CERs.

The part of the process which distinguishes the commercial parametric model from the building of spreadsheet models is the final part – continuous calibration. This ensures that the commercial cost model is maintained. While spreadsheet models tend to be created for a project and then not re-used, commercial models are continuously used and need to be maintained. In parametrics, the algorithms or Cost Estimating Relationships are recycled, thus ensuring a return on investment for the time spent researching.

Figure 2.2 is an example of a simple parametric cost model. In the example, it was established that on average the door took 30 minutes to prepare and for this configuration it took 60 minutes to paint. The project characteristics are provided to go with this data: the area, number of layers of paint and the number of sides painted. As is clear to see, a simple equation can be produced to relate the independent variables (characteristics) to the dependent variable (time taken). The parametric model can be re-used for different scenarios and different projects.

This is a core equation. Keep in mind the fact that we do not take all the parameters which could impact the cost – such as the brush size, work regulation, skill of the painter and so on – into account. Commercial parametric cost models take many person years of cost research. They involve many cost estimating relationships and numerous Cost Drivers. Parametric cost models are conceived in the same manner as the simple door painting model, that is, built around a core equation. In the case of the PRICE Systems hardware model, the core equation is the weight and technology index.

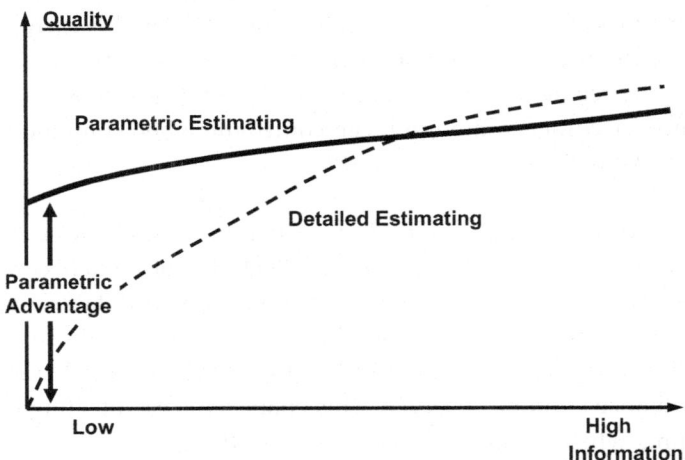

Figure 2.1 The advantage of parametrics

Characteristics	Observations
Door area	2 square metres
Layers of paint	2
Sides painted	2

Characteristics	Metric
Preparation time	30 minutes
Painting time	60 minutes

Equation:

Time (minutes) = 30 + 7.5 × (Area × Number of Layers × Number of Sides)

Figure 2.2 Parametric model example

Parametric vendors will employ a Chief Scientist and a cost research team. This team will be well versed in statistics, mathematics and cost research. They generate the parametric cost models which are communicated to the software programmers in the form of a White Paper. These White Papers are the documents which link the cost research to the software models. They enable commercial organizations, who supply cost models, to employ the best cost analyst for cost research and expert programmers for the software models, without the two roles complicating the two separate disciplines.

It is often said that if cost research could be provided in tablet form life would be simpler, however, for the time being software is the easiest way to communicate cost research.

Use of Parametrics throughout the Organization

Commercial parametric models have many applications (see Figure 2.3). At the top of the organization, they are able to solve problems for the decision-makers and senior management. At this level, senior managers can more quickly consider costs which influence the decision. Hence, bid or no-bid decisions can be made with a view to the cost. Furthermore, market strategy and business plans can be assessed alongside benchmarked productivity, competitor analysis and productivity tracking.

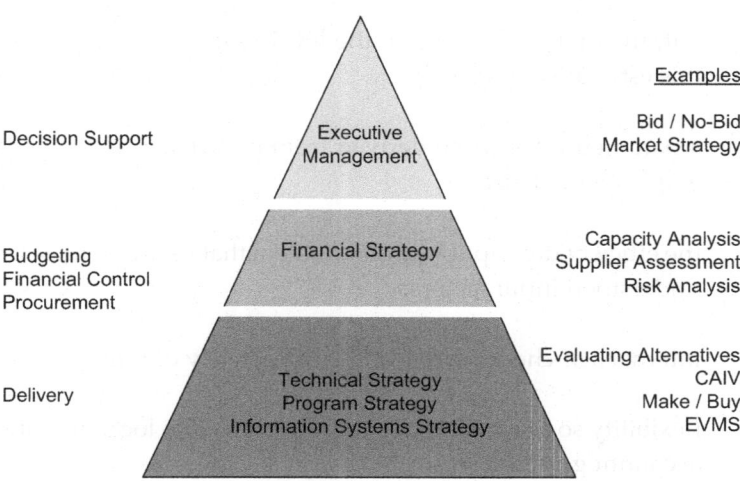

Figure 2.3   Enterprise applications

At the executive level, financial decisions can be accelerated with parametric estimates. Capacity planning will ensure best use of the available resources, supplier assessment employed to select the preferred supplier, and risk analysis used to consider the uncertainty in budgets and financial reporting.

At the operational level, parametrics can be deployed to deliver technology programs on the basis of the effective consideration of alternative options and the tracking of the program throughout its life with Cost as An Independent Variable (CAIV) and Earned Value Management (EVM) (through the integration of parametric estimating).

Ease of Use

Commercial parametric models, such as computerized models, offer many benefits over manual systems: reduced risk of accounting mistakes, faster operation, added capability, and greater facility for sensitivity analysis. The essentials of operating commercial cost models are easy to learn and provide a fast and effective method of cost estimating. The key ingredients of this methodology are:

- interactive operation with easily identifiable inputs which prompt answers through a dialogue with the user, for example, 'Percentage of new design?';

- outputs that are defined to the level of the tasks conducted when this estimate in finished;

- a parametric approach derived from experience and supported by empirical evidence;

- the efficient description of the problem that uses a small set of easily understood input factors;

- internal self-checking to test the consistency of input data sets;

- flexibility so that the model may be adapted to local definitions and accounting procedures;

- performance calibration that relates current estimates to actual achievements on prior projects.

The Windows-based user interface provides control of the model, its outputs and the storage of input data. The simple interface allows the user to develop and re-use large numbers of historical files. It enables the parametric model to be used following a course of intense instruction, and provides an audit trail for any study. An illustration of the look and feel of a parametric cost model is provided in Figure 2.4 with an example of a GPS Receiver system estimate.

In December 1995 the Commander of the Defence Contract Management Agency (DCMA) and the Director of the Defence Contract Audit Agency (DCAA) in the United States sponsored the Parametric Estimating Reinvention Laboratory under the Parametric Cost Estimating Initiative (PCEI). The purpose of the Reinvention Laboratory was to test the use of parametric estimating techniques on proposals and to recommend processes to enable others to implement these techniques.

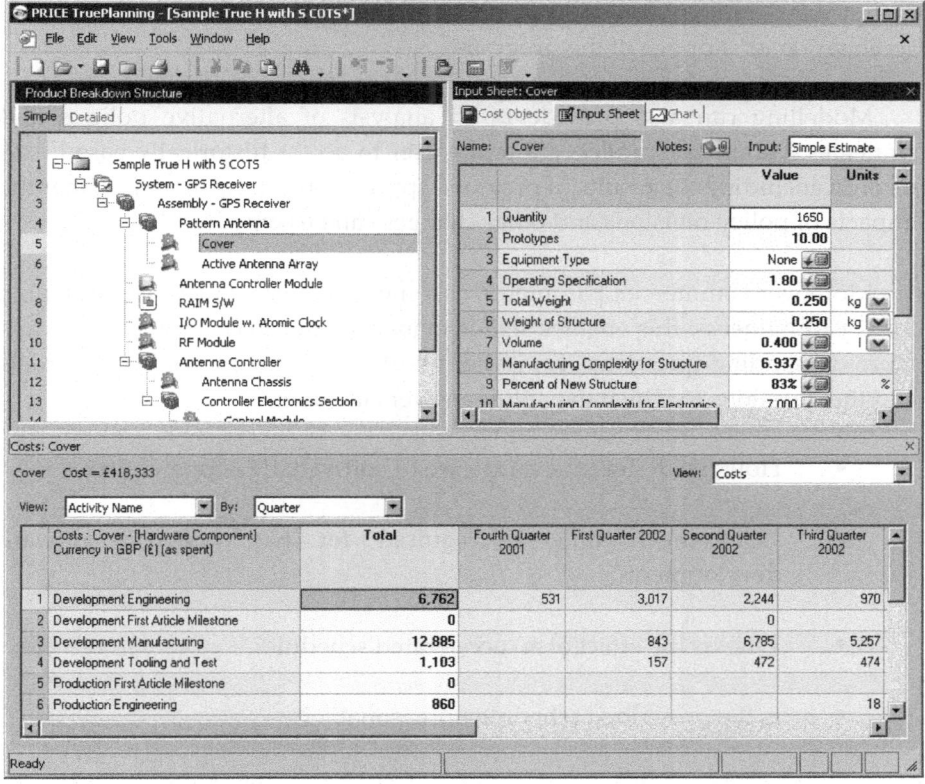

Figure 2.4   Example parametric cost model – Product Breakdown Structure (PBS), Cost Drivers and outputs

Thirteen Parametric Estimating Reinvention Laboratory teams tested and implemented parametric techniques. The estimates covered the range of use from specific elements of cost to major-assembly costs. The teams generally found that using parametric techniques facilitated rapid development of more reliable estimates while establishing a sound basis for estimating and negotiation. In addition, the teams reported proposal preparation, evaluation and negotiation cost savings of up to 80 per cent, and a cycle time reduced by up to 80 per cent.[1]

A parametric model permits rapid and early probability cost evaluations based on project scope, program composition, and demonstrated organizational performance. Operational and testing requirements are incorporated, together with technology growth and inflation.

In addition to cost, parametric models derive typical schedules for the work to be accomplished. Schedule constraints that have been imposed are examined within the parametric model, and costs are adjusted to account for apparent acceleration or slippage of the program.

Modelling can also provide rapid analysis of alternative policies and strategies. Parametric models allow the user to access historically based data records, including the results of previous applications of the model, so that the impacts of policy revisions and other changes can be compared.

Moreover commercial parametric cost models offer a quick response time for Cost Engineers; this enables senior managers and decision-makers to ask more demanding questions. Typical applications of commercial cost models and the questions to which they offer answers include:

- How much does the hardware or software development cost?

- How much time is required for hardware or software development?

- What is the effect of an accelerated schedule?

- Is the prescribed schedule unreasonable?

---

1 Chapter 1 – *Introduction, International Society of Parametric Analysts (ISPA) Parametric Handbook*, Issue 2.

- What are the trade-offs between schedule and cost?

- What size hardware or software development project can be undertaken with a fixed budget?

- What are the cost impacts from varying efficiency, experience, or design skill?

- How will costs vary among different organizations?

- What are the effects of the operating environment and hardware reliability requirements?

- What savings can be obtained by using customer-furnished or existing hardware or software?

- How much does it cost to modify existing hardware or software?

- How much will technology growth affect hardware or software development?

- What effect do modern hardware and software development practices have on cost and schedule?

- How does previous experience relate to new projects?

The Complexity Concept

A commercial parametric estimating tool will have the ability to adapt itself to many different environments, both business and industry. Parametric models also have the ability to represent the magnitude of a problem as a simple core equation, adjusted using secondary equations. In some parametric models these inputs are hidden from the user, in others the numeric inputs are explicit.

Cost models also need a basic hypothesis for this core theory. The simplest theory for a hardware parametric model, for example, is the higher the weight, the higher the cost, for the same technology and productivity. Engineering instinct indicates that this holds true for all situations, providing the technology and productivity are the same. If either of these parameters changes, then all bets are off. These technology and productivity elements are encapsulated in a single input parameter for hardware called Manufacturing Complexity. But in the case of software cost models several separate parameters are often involved. This core theory forms the central equation of many cost models.

In simple terms the central equation finds that the more of a single type of technology acquired, the more it will cost (see Figure 2.5). In the instance of hardware, this magnitude of technology is commonly assessed in terms of weight. Hence, the larger the item (for example, an aircraft wing) becomes, the greater the expense required to manufacture it in terms of the material content and labour. If the wing needs to be lighter, this can be achieved by changing the technology, but this may not necessarily result in lower cost as the new Manufacturing Complexity of the lighter technology is likely to be higher – resulting in a high normalized cost.

If it is accepted that cost is a function of the weight and the complexity, it is possible to obtain a normalized basic cost. To estimate all scenarios, several additional parameters need to be added, through the adjustment equations, involving elements such as quality, quantity, skill of the team, and so on to achieve a realistic estimate. Using the simple door-painting example, adjustment equations are the equivalent of the brush size, labour experience and so on.

The effect of producing a normalized cost density from the historical projects is to generate a trend line or average Manufacturing Complexity on condition that the technology involved is the same.

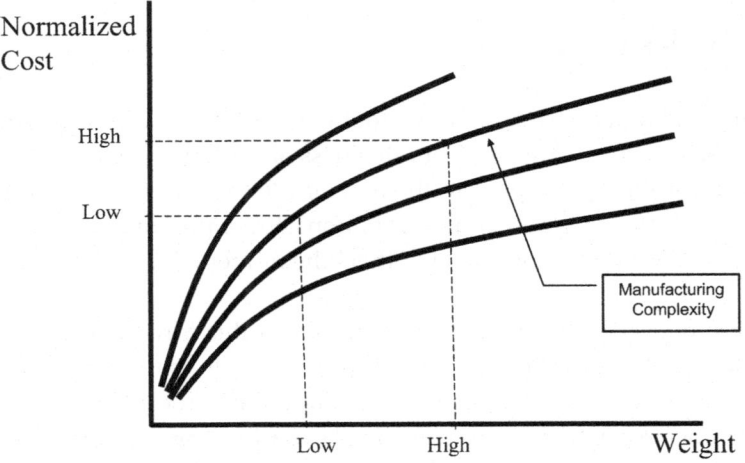

Figure 2.5 **Normalized cost as a function of weight and complexity**

MANUFACTURING COMPLEXITY

The complexity factor is the normalized representation of the cost density; though it is not measured in dollars ($) per pound (lb) or Euros (€) per kilogram (kg), these units provide an easy visualization of the parameter. In fact it is a unit-less, empirical factor comprising the technology implicit in a product and the productivity of its manufacturer (see Glossary).

By way of illustration, consider something heavy and inexpensive, such as a railway rail, for which the Manufacturing Complexity factor is low. Compare this to something that is light and yet expensive such as a hybrid module for a satellite, where the Manufacturing Complexity factor is high.

Following the calibration process, historical project information regarding equipments, sub-systems or systems will have been stripped of their unique project features. A process of normalization will have eliminated the effects of differences in quantity, economics, production rate, technology maturity and so forth. Mathematically the residual at the end of this process is a representation of the technology, from which the item was produced, and the productivity difference between one producer of the technology and another.

In the example in Figure 2.6, the 1960s Harrier has been in existence for a considerable time and needs to be replaced by a new generation of Joint Strike Fighters (JSF) with Vertical Short Take Off and Landing (V/STOL) capability. The next generation of aircraft will have greater capability provided by newer technology, placing it further along the technology axis.

The F16 has been produced in many variants since the late 1970s, hence it is placed between the Harrier and the JSF in terms of technology. But it has been manufactured both by the original systems designer and other manufacturers. Hence the technology will be constant, but the productivity of the company producing it will result in a different Manufacturing Complexity. The company with the lower Manufacturing Complexity will produce the F16 for the lowest cost due to their higher productivity.

HOW ARE COMPLEXITY FACTORS OBTAINED?

There are several methods for obtaining complexity factors, the simplest of which involve embedded tables in the models. These calibration tables contain typical values for a product or technology: they contain reference (industry

Figure 2.6   Manufacturing Complexity attributes
Source: http://www.navy.mil/search/photolist.asp

average) complexity factors based on data collected over a long period. These factors can be used to generate a 'should cost' when internal data is missing.

It is also possible to derive parameter inputs via calculations, for example, Manufacturing Complexity generators derived from an assessment of the technical characteristics of the equipment. For mechanical Manufacturing Complexities these characteristics include:

- material;

- process;

- precision or tolerance;

- number of parts or layers of composite;

- hogout (the manufacturing process of rapid material removal when machining from solid);

- roughness.

For electronics the generator requires inputs such as:

- type of electronics;
- type of components;
- quality level;
- density level.

These characteristics are the type of factors a traditional estimator would intuitively assess when conducting a detailed estimate of an item. Some parametric models only provide access to the calculator inputs and not the technology factor itself; the complexity factor is buried within the model.

KNOWLEDGE BASES

A common approach to retaining project knowledge is through the use of a knowledge base. This is the third method of obtaining and storing parameters for the costing models. The cost knowledge database (which ideally is hosted on the vendor's web site) contains typical model inputs for the systems that are to be estimated. These are based on research gleaned from multiple historic projects and provide a vehicle to deliver up-to-the-minute cost research data to the user, for example, data from the NASA/Air Force Cost Model (NAFCOM) database. These knowledge databases are effectively search engines which filter the data and provide the results of the selected criteria quickly and easily.

It is also possible to have a knowledge database including information developed with your organization; this database is held centrally or located on a local machine to provide input parameters for estimates.

Finally, if all other sources of ready-to-use Manufacturing Complexity are exhausted, or if the technology is unique and cannot be found in the sources referred to previously, a technique of calibration can be employed. Data sources such as tables and knowledge bases will provide average industry Manufacturing Complexity, but calibration, with its sources in an organization's own prior projects, will provide a Manufacturing Complexity *that reflects that organization's own productivity as well as its own technology*. This makes for a time-consuming, but ultimately most accurate, approach.

Cost Modelling with Complexity

The detail of how to estimate using parametrics is the subject of the next chapter. Let us first examine the central role of the Manufacturing Complexity factor or normalized cost density.

An alternative representation of a parametric model is shown in Figure 2.7. When calibration takes place, historical projects are subjected to analysis and scrutiny. The parameters of the costs model are input into the model and the historic project cost is filtered through the layers of the model until it is output as a normalized cost density or Manufacturing Complexity. This filtering process is fundamental if these parameters are to be used as a starting point for future estimates. By populating this parameter at the core of the model, similar projects can be estimated. Providing the Cost Engineer is consistent in his or her use of the model, the mathematics will apply equally forwards, as well as backwards.

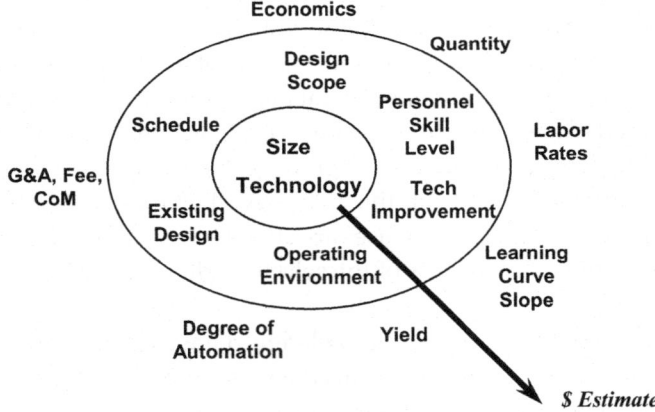

Figure 2.7 Simple representation of a parametric hardware model

There is nothing wrong with estimating by analogy and other techniques. However, sometimes these techniques need the assistance of a little science. Take the example of an estimate in Figure 2.8. This is a straightforward estimating task for many Cost Engineers.

Using analogous estimating techniques it is easy to determine the cost of the new Motor Control Card (MCC) on the basis of the available historical data and the fact that the cards are all from the same family of technology.

Product Family: motor control card (MCC)

MCC1		MCC2		MCC3		MCC4		MCC5	
weight	0.6	weight	0.4	weight	0.7	weight	0.35	weight	0.55
Quantity	35	Quantity	20	Quantity	75	Quantity	12	Quantity	150
schedule	6/96-11/96	schedule	3/97-9/97	schedule	5/98-12/99	schedule	3/99-8/99	schedule	5/99-8/00
Prod rate	7	Prod rate	3	Prod rate	4	Prod rate	2	Prod rate	15
Cost	15	Cost	18	Cost	12	Cost	10	Cost	13

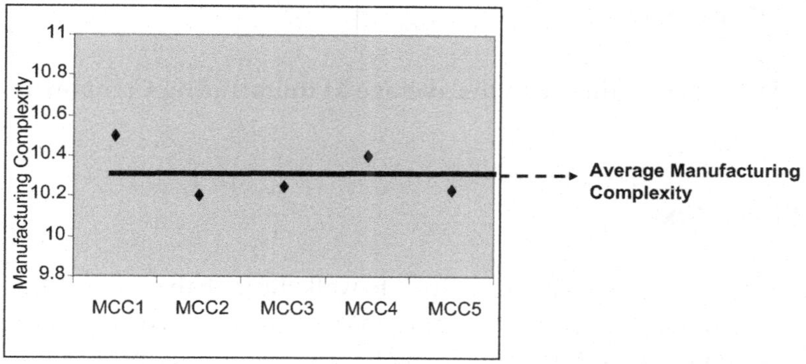

Figure 2.8 Determining an average Manufacturing Complexity

The new card weighs 0.65kg, by analogy MCC1 was a similar weight; hence we can derive that the new card should cost about €15k (the same as MCC1). But on closer inspection only 35 of MCC1 were made and the new card will have a production quantity of 110. The last column shows that 150 of MCC5 were produced for €13k. But in that case the production rate was 15 per month and the new card will be made at a rate of six per month.

In other words, even with the detailed information available, this is a complex problem. A good Cost Engineer should eventually be able to produce and justify their analogous estimate. However, with parametrics the process is easier. All the projects will be calibrated, run backwards through the model and the complexity obtained.

Having determined the Manufacturing Complexities for the historical Motor Control Cards it is possible to calculate an average Manufacturing Complexity for this family of technology. The average Manufacturing Complexity represents the technology index, the scatter around the average is due to the productivity of the organizations that produced the individual items, some better than average and some worse. The average manufacturing complexity can be run forward through the cost model, with the new MMC characteristics (see Figure 2.9), to determine the estimated cost.

Product Family: Motor Control Card (MCC)

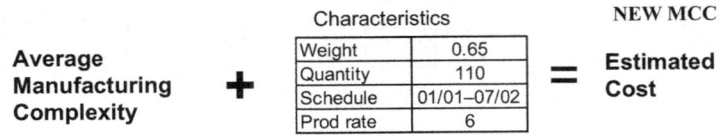

Figure 2.9   Estimating with the average Manufacturing Complexity

The 'Black Box'

The easiest way to explain parametrics is to walk through the construction of the model. This should provide reassurance that parametrics is no more a mystical 'black box' model than Microsoft Excel. Figure 2.10 is a simple example of an Excel formula: SUM. When 2 is entered in cell B:2 and also in cell B:3, Excel provides the correct answer of 4 (in B4). The sum of 2 and 2 is 4. But what does =SUM(B2:B3) mean? The point is, we see the effect. When we use Excel there are large amounts of code running in the background that perform the requisite mathematics. The outcome is the desired result, 2+2=4, but the figure 4 is the result of internal code.

First generation mainframe and second generation personal computer parametric models are similar to 'black box' models: commercial models that include Intellectual Property (IP) equations within them. Some of these are simple equations, learning curves, curves to reflect market forces on technology maturity, schedule effects cost curves and more; all of which have an expected influence on the outcome. Other algorithms are trade secrets, but the analogy remains. With first generation parametric models, a result is seen that can be explained, just as in the Excel spreadsheet. These equations result from cost and schedule research.

The fundamental characteristic of the parametric inputs is that of interrelationship. A change in any one parameter is usually not localized to one cost element, but rather, may have a direct effect on several cost elements, and an indirect effect on many more. Consider the impact of a change in quantity. Certainly this will result in a change in manufacturing cost. But it might also affect the fabrication process and, hence, the cost of tooling equipment. In addition, a change in quantity will probably have a schedule effect and as a

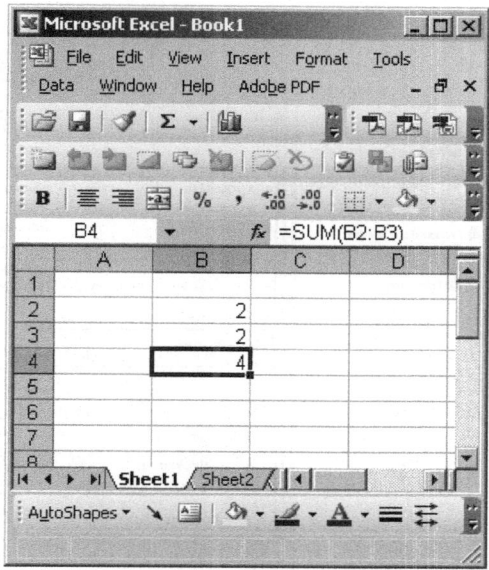

Figure 2.10 SUM formula in Excel

result place the cost in a different time period where inflation effects would change the costs through application of different escalation of the costs. A filtered impact on integration and testing, sustaining engineering, and project management would almost certainly result from a change in quantity. This dynamic effect, involving multiple impacts, is characteristic of most input variables.

Parametric models contain thousands of mathematical equations relating the input variables to cost. Each specific set of input parameters uniquely defines the hardware, software or information technology (IT) for cost modelling. The resultant cost output is determined from the mathematical equations alone. Most importantly, parametric models do not perform the function of a table look-up model and are not influenced by optimism or pessimism.

The third generation parametric model arrived with the new millennium. This model involves open architecture within which the algorithms are visible (see Figure 2.11) when viewed in the software application. Consequently, third generation parametric cost models are easy to validate and verify.

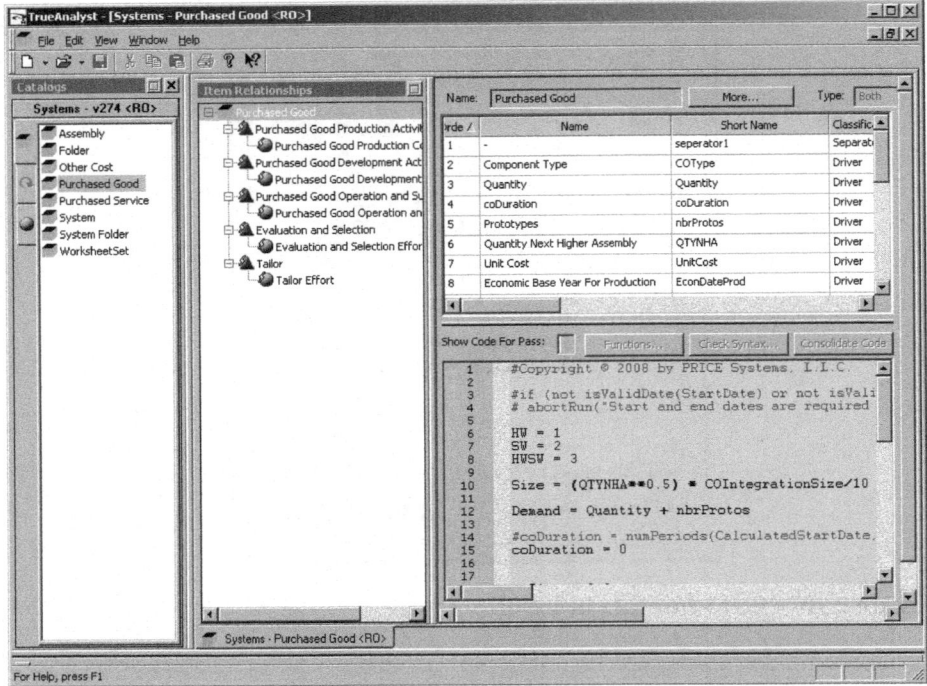

Figure 2.11 TrueAnalyst from PRICE Systems open architecture with visibility of CERs

Most parametric models have been designed to estimate costs with a minimal amount of software, IT or hardware information. This feature makes them a legitimate tool for program cost estimation during the conceptual stage of program development, since the model can use its default internally generated values for any missing input variables in order to estimate cost. Of course, it is always preferable for the subject matter expert to supply the inputs, when their values are known. In this way, the statistical uncertainty of the parametric model is reduced.

The characteristics of a third generation parametric cost model are:

- distributed access to the estimate through a genuine client/server architecture;

- the application of Activity Based Costing philosophy to structure the estimate;

- the inclusion of hardware, IT and software in one breakdown structure with management oversight and integration and test costing included through cost model interoperability;

- an open framework in which cost models can be built.

Summary

The lessons learnt in this chapter will have enabled you to appreciate parametrics through the following:

- Parametrics provides speed in estimating cost and schedule when little information exists regarding a project, proposed program, a competitors' product or a suppliers' equipment.

- Parametric estimating can be considered similar in many ways to estimates constructed in spreadsheets by cost engineers.

- Commercial computerized parametric models offer many benefits over manual systems including 80 per cent saving in cost and time.

- Parametric models have been designed to estimate cost and schedule with a minimal amount of information.

- The normalized cost density of a technology, also referred to as manufacturing complexity, can be determined through tables, calculators, knowledge bases and calibration.

- The latest third generation parametric models are open cost estimating frameworks which can accommodate proprietary models or bespoke models which can be easily validated and verified.

3

How to ... Estimate Using Parametrics

The parametric estimating process is both simple and logical. Part of its beauty is the top-down nature of the approach. The model is not derived from the bottom-up approach which would have the potential for cost duplication and omissions. This is because the process is driven by determining the magnitude of the problem: for hardware the magnitude of the technology and the organization's productivity and for software the organization productivity and size. You don't enter costs, non-cost data is entered and costs are created. The process for estimating cost and schedule involves four simple stages:

1. determine the system architecture in a breakdown structure;

2. identify the project parameters for the elements of the system;

3. determine the cost and schedule drivers of the elements of the system;

4. run the model, validate the results and present the answers.

The Parametric Modelling Process

The first step in the process is to identify the Estimating or Product Breakdown Structure (EBS or PBS).

As indicated in Figure 3.1, with this mobile phone the PBS cannot be conducted in total isolation. One of the great benefits of this estimating technique is that it forces the whole team involved in the estimating process to communicate. They all need to understand the architecture of the system to be

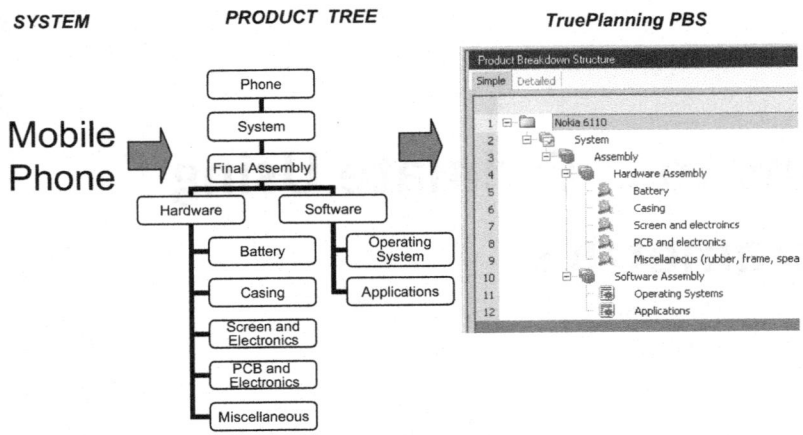

Figure 3.1 First step – estimating

estimated. This can be discussed and represented as a structure directly in the cost model system.

The Graphical User Interface (GUI) is such that building the Product Breakdown Structure (PBS) is simple. The emphasis with parametrics is on the Cost Engineer spending more time talking to the engineers, not building the cost system.

When deriving a Product Breakdown Structure (PBS) for an estimate it is easy to construct it, but there are some important points to consider. It is important to ensure that the technologies within the item are the focus, bringing out the electronics and the types of structure – for example sheet metal, machined or composite. Anywhere there is electronics there is the potential for software, so consideration needs to be given to what software is present and into which electronic items it is integrated. Thought also needs to be given to the procurement philosophy of the project, which hardware and software the organization is going to design and build in-house and which is going to be purchased or procured to specification. This leads to the integration and testing of the technologies. Finally, consideration must be given to what is the natural flow for the items to be built into sub-assemblies and these sub-assemblies to be built into assemblies, before becoming systems. This will enable the PBS to be constructed in a meaningful and logical engineering structure.

Once the architecture has been captured, it is possible to enter the first level of parameters into the model. These first level parameters are designed to be available from systems engineers even at the early stages of the project. As can be seen in Figure 3.2, these parameters can be common to all the items in the breakdown structure, for example quantity of production items, thus reducing the number of inputs required in each element.

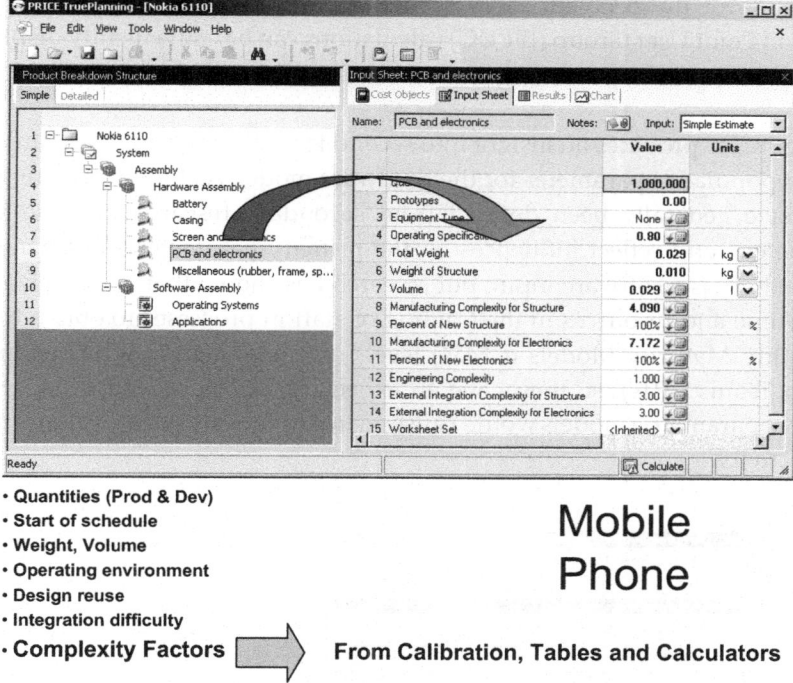

Figure 3.2   Second step – first level parameters

Ideally all the inputs should be viewed on a single page to enable the whole product to be seen at a glance. It is also easier to use the tool if it is grouped into different input parameter types, for example, software size parameters, schedule parameters, risk parameters, and so on.

One of the fundamental difficulties of parametric estimating is determining the magnitude of the problem to be estimated. Having determined the technology involved, in hardware terms, all that remains is to establish the weight of electronics or structure. In terms of software, determining the magnitude of the problem involves consideration of the Source Lines of Code (SLOC), Function Points (FP), Predictive Object Points (POPs), Use Case Conversion Points (UCCPs) or Functional Size (for Commercial Off the Shelf [COTS] components) (software sizing is covered in detail in Chapter 12). Having determined the size of the project in these terms, the challenge has been significantly reduced. What remains is consideration of the program and industrial environment.

Parametric models contain embedded calculators to enable assessing the magnitude of the problem more accurately. For example, the International Function Point User Group (IFPUG) calculations can be conveniently displayed in the software to help in sizing it.

Figure 3.3 provides an insight into second level parameters. These are used to help populate a parameter for the hardware model. As their name suggests, they have generally been derived from secondary research into the input parameters, and reflect influences on the primary parameter. Organizational Productivity is a software input, but research has shown that as organizations invest time and resources in the training of staff in processes to obtain higher Capability Maturity Models (CMM) levels and team-building for Integrated Project Teams (IPTs), so they have an expectation of improved productivity, which in parametric models would cause an improvement in the Organizational Productivity.

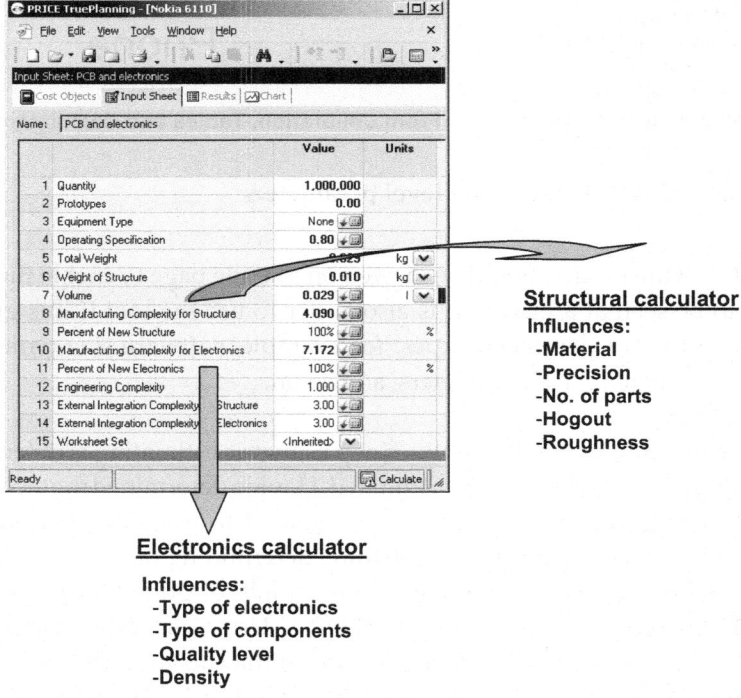

Figure 3.3   Third step – second level calculator inputs

HOW TO ... ESTIMATE USING PARAMETRICS

It is important that the outputs of the model are validated for reasonableness. Just because the estimate is generated by a computer using mathematical relationships there is no excuse for not validating the result. A good parametric model will validate the inputs and provide supplementary information to help check the outputs.

Figure 3.4 demonstrates the outputs from a commercial parametric model.

One aspect of any commercial estimating model is the consistency achieved from a professionally built Graphical User Interface (GUI). The attraction of a basic spreadsheet is the flexibility of the application; however, this flexibility is also its weakness. You may be able to create any spreadsheet you wish: accounts received, cash flow, blood pressure tracking, blood sugar chart, soccer tables, and so on, but you are unlikely to be consistent in how you do so. In which cells of the spreadsheet do you type the input parameters? In which cells of the spreadsheet will you find the results of the model?

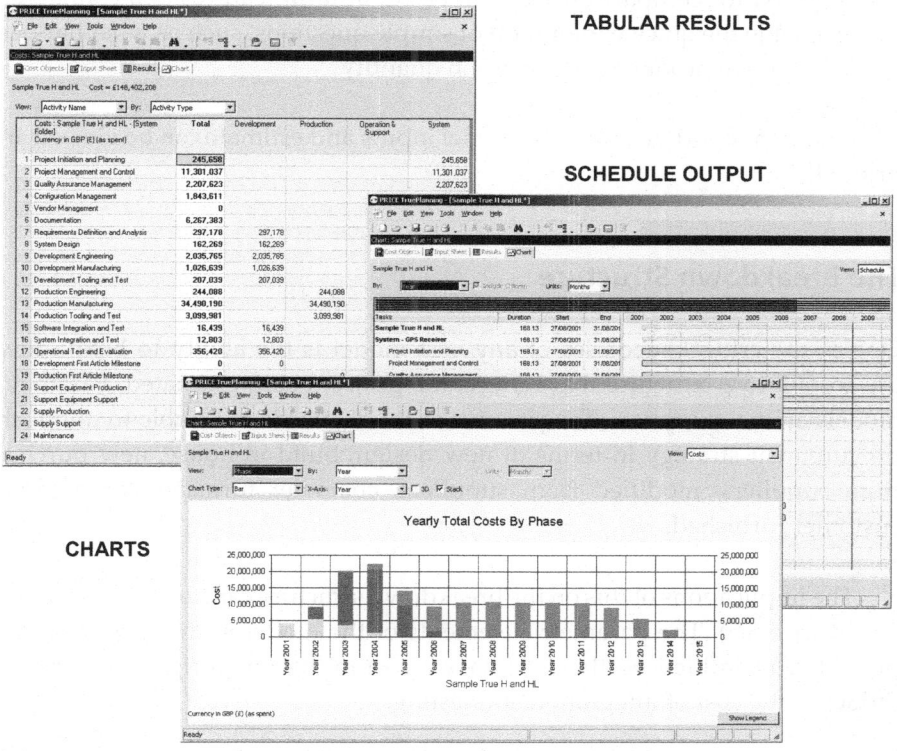

Figure 3.4   Run the model and outputs

The commercial model is less flexible, but more efficient. The Cost Engineer knows the Cost Drivers that need to be collected, and the Cost Engineer can spend time and effort gathering the most accurate information to populate the Cost Drivers. The Cost Engineer, using a spreadsheet, needs to create a parametric model and determine the best Cost Drivers to influence that parametric model; only once the model is prepared can the Cost Engineer consider producing the estimate and gathering information for that purpose.

Third generation parametric models provide the best of both worlds – a cost modelling capability which has been designed for Cost Engineers including either commercial models or organic, home-grown models in a single framework.

Once the hardware, software and IT project have been described, the parametric model will estimate costs and schedules. Numerous bar and pie charts are available to view the model's estimate in graphic form with features that enable the creation of a baseline estimate against which the estimate can be re-computed for comparison. These types of built-in features of a commercial model enable the quick exploration of simple questions, such as what happens to the cost if we produce an increase in quantity?

Finally, a report detailing the model inputs and estimate can be viewed and printed.

The Breakdown Structure

One of the critical successes of any cost model is the ability to break down the costs into a structure that reflects the project being estimated. Parametric models are no exception. The parametric model needs to be able to mimic the procurement strategy in terms of new design, build in-house, new purchase from suppliers, modified from suppliers, customer furnished or modified customer furnished.

The implications of this on the breakdown structure can be easily visualized (see Figure 3.5). The parametric elements of the model in the table have two possible cost influences: they can have a cost in their own right and they can influence the cost of integration and testing.

So if the organization is going to design and build an item of hardware or software, the item itself will have development and production costs, but when

HOW TO ... ESTIMATE USING PARAMETRICS

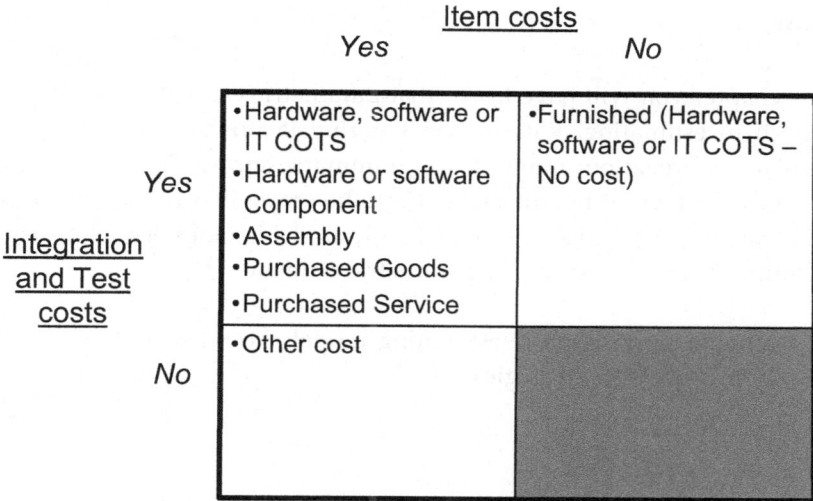

Figure 3.5   Elements in a parametric model

offered up to another items in the system, it will need to have an influence on the integration and testing. Similarly, if the item is purchased, then the item is paid for and has an influence on integration and test costs.

Furnished items are too often seen as free when provided by a customer to a supplier. However, whilst they may not have a cost in their own right, they need to be included to ensure that the costs of integration and testing are considered. They will also need to be designed into the system to ensure that they are properly accommodated once they are integrated. It is too late once these items have been delivered to find that they require a 32-bit signal rather than a 64-bit signal, or 40 volts power and not 12 volts. The same can be said for purchased items.

Modifications to furnished or purchased items also need to be considered. These are presumably low in cost, relative to a new build; otherwise the item would not have been purchased.

Finally there are the other costs; these have a cost in their own right, but have no integration and test influences. For example, the acquisition of a new hanger, dry dock or production facility might be a legitimate cost to the project, but the integration influence is not a consideration.

Summary

In this chapter you will have learnt that parametric estimating is a top-down approach to estimating. Rather than entering labour and material into the system for aggregation; non-cost programmatic and design parameters are entered for cost and schedule estimates to be created. Using a spreadsheet is flexible, but a commercial third generation model is more efficient. There are four simple steps to producing a parametric estimate.

A commercial parametric estimating model should be able to mimic all necessary procurement strategies.

4

How to ... Prepare Bids Faster with Fewer Resources

Without a doubt the most sought after attribute of parametric estimating is the speed without any loss of accuracy.

It is common practice to prepare a bid in a defined time scale. The request for a proposal arrives and a submission date and time is included. A kick-off meeting occurs and, with more reviews of the request, a target cost or the customer's known budget is established. These establish the dimensions of the problem.

With an analytical detailed, bottom-up or grass-roots methodology each work package manager is asked for an opinion of the cost of their work package. At the same time they are asked for a work package description, risk assessment, technical text, and so on. As such, a detailed estimate is slow to be produced. The engineers' estimates follow behind the work package descriptions and the compilation and checking of all the engineers' estimates is tedious and labour intensive.

As can be seen in Figure 4.1, the total cost estimate is commonly only known a few days before the submission date. What ensues is a reactive culling of the figures. Exclusions and assumptions are made in order to reduce the costs of this without justification and bad decisions result.

Bid/No-Bid Decision

With parametrics it is possible to determine the first estimate very quickly. It should form part of the bid / no-bid decision-making process. This first estimate will not necessarily be as accurate as the subsequent detailed estimate, but it

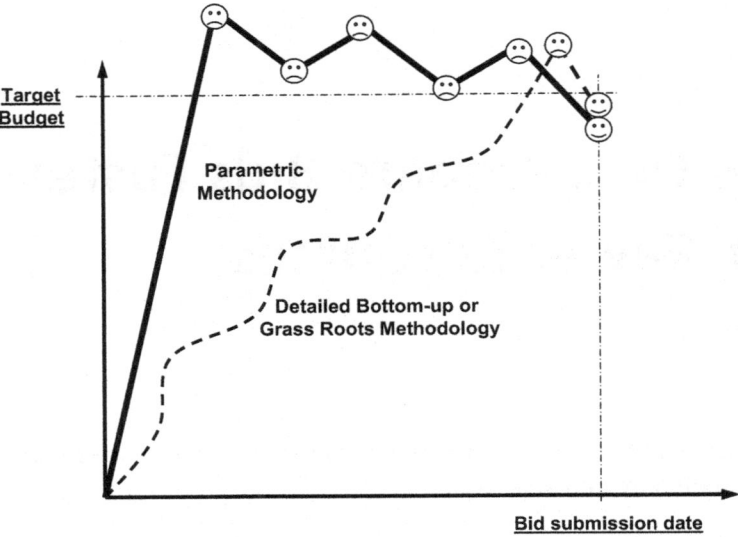

Figure 4.1   Rapid estimating with parametrics

will enable decision-makers to be proactive and decide in a logical manner what to do.

Generating a proposal can take several years, with a considerable number of changes and regular staff turnover it is important to avoid starting again from scratch with each change, and to ensure continuity and consistency in the use of estimating tools and bidding methods.

Parametric tools enable an efficient management of various proposals. Figure 4.2 illustrates the concept of repeated designs that are cost estimated, each iteration involving a refinement of the design and the estimate. Due to the length of time it takes to generate a detailed cost estimate, the only practical way to achieve this structured development process during a bid is by adopting parametrics.

Schedule Estimating

One of the unique aspects of parametric models is that the algorithms are not limited to cost. The equations in the models provide not only Cost Estimating Relationships (CERs), but also Schedule Estimating Relationships (SERs). It is possible to determine an independent view of the schedule. Research

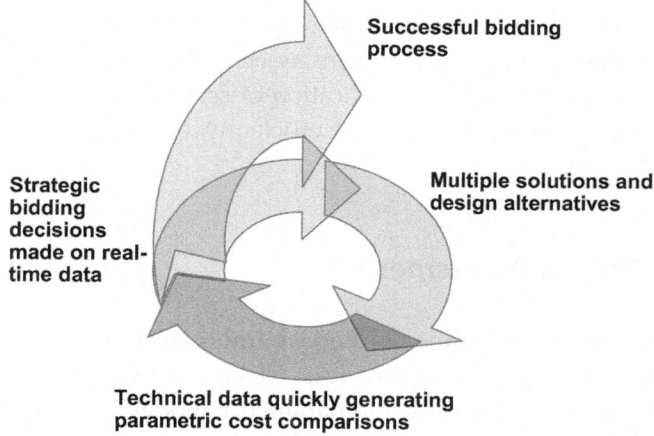

Figure 4.2   Iterative proposal preparation process

techniques that produce equations that drive cost and schedule can also determine reliability measured in Mean Time Between Failures (MTBF), which can be used in the calculation of the corrective maintenance cost. When the MTBF occurs, this will trigger expenditure in spares and repair resources.

When input parameters are changed, the schedule will recalculate and reflect the associated modification to the program. Thus the skilled Cost Engineer can monitor the implications of technical change on both the schedule and the costs. This is a situation with which many Cost Engineers will be familiar; one in which the eternal triangle of technical performance, schedule and cost are linked. If any one of these elements is changed, there is an influence on the other two. However, in traditional analytical estimating it is difficult to track these changes quickly.

When combined with a parametric framework the system can scan across multiple projects and monitor their interrelationships. Using an Integrated Master Schedule enables the provision of advice to senior staff on industry capacity. Industry capacity is the measure of manufacturing capability of different industrial segments. This is interesting to procurement agencies: the UK Ministry of Defence (MoD) will consider the total labour available in the UK shipyards, for example, and try to stagger the demand on shipbuilders to ensure that their capacity is not exceeded. This ensures that they will not invoke an excessive premium either for training new staff during a trough or dip in workload or for overtime payments during a peak in workload.

This is particularly important when dealing with Government Furnished items (GFX), that is, when Government assets are being held for use in trials or for other project needs. Early indication of schedule slippage is important to ensure that those assets are used efficiently or released for operational deployment.

Schedule Penalty Assessment

The accuracy of the schedule forecast is important since labour costs are significantly influenced by the schedule. Given a development schedule and appropriate engineering inputs, the parametric model calculates engineering costs and a typical schedule.

If the Engineering Complexity of a project suggests more effort is required than is normal for the schedule, a cost penalty is assessed to reflect the accelerated schedule. Conversely, costs for program stretch will be included, if the complexity of a project suggests a shorter schedule than that specified. If schedule acceleration becomes excessive, typically the model will provide a warning message that will prompt the questioning of inputs. Figure 4.3 shows the built-in bathtub curve typical of parametric models.

The period of the schedule over which actual delivery of production hardware takes place is also important. The length of this period is likely to determine the Monthly Production Rate, the amount of refurbishment for special tooling and test sets, the percentage of Engineering Change Notices (ECN), and the impact on Project Management. The schedule calculation assesses penalties for acceleration or stretch-out when the schedule entered exceeds the calculated production schedule length of the model. This schedule estimating capability provides an Earned Schedule capability not realized in many Earned Value Management (EVM) systems.

Parametric models thus support whole-life costing by integrating the estimate of cost and duration of projects. They calculate and report resource, activity, component and system level costs. They also calculate and report resource, activity, component and system level effort and duration, thus providing comprehensive understanding of project cost and schedule requirements for faster bid preparation.

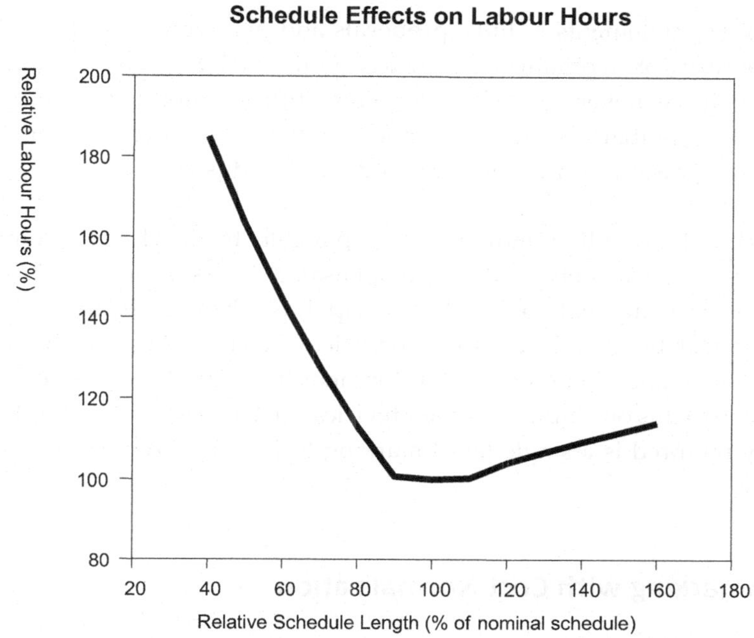

Schedule Effects on Labour Hours

Figure 4.3   Modelling schedule penalties

Anecdotal Evidence or Proof in the Bid?

In an interview in *The Times* newspaper[1] the retiring UK Chief of Defence Procurement, Sir Robert Walmsley, stated, 'I think we have an obligation to work with industry to ensure that our suppliers do not just remain world class in defence, but aspire to be world-class manufacturers that *can withstand comparison to other industries*.'

This repeated a statement he had earlier made when addressing the Public Accounts Committee:

> ... we do try to do benchmarking and we have been trying to get going a benchmarking study on ship pricing ... by relating it to ... complexity of a warship to the relative simplicity of a merchant ship ... We are always looking for better 'Should cost' data.[2]

---

1   *The Times*, 7 June 2002 (emphasis added).
2   Select Committee on Public Accounts Minutes of Evidence, Examination of Witnesses (Questions 20–39), Wednesday 14 November 2001, Mr Kevin Tebbit, Sir Robert Walmsley

Many of the statements in bids, proposals and marketing literature claiming the superiority of an organization are based on anecdotal, or at best qualitative, evidence: it has never been possible to quantify the productivity of industry to demonstrate that one nation is more productive than another or that one manufacturer is more efficient compared across industry.

In the shipbuilding industry it is possible to benchmark commercial shipyards using a metric called Compensated Gross Tonnage (CGT). CGT measures the complexity of building a ship. It is calculated by multiplying the gross tonnage by the relevant CGT coefficient (calculated by the Organization for Economic and Commercial Development [OECD]). However, this is only applicable to this one industry. Other metrics exist in other industries, but what is really required is a single benchmarking technique and measure across all industries.

Benchmarking with Cost Normalization

As established in Chapter 2, Manufacturing Complexity represents the normalized cost density of an item that has been manufactured. It has two dimensions: technology and productivity. Hence, more than one Manufacturing Complexity can exist. As aircraft are made using different technologies and by different companies, these have various complexity figures. However, if the technology is constant, the only dimension that changes is productivity. As a result, it is possible to compare the efficiency of companies that produce similar technologies.

SUPPLIER ASSESSMENT

One obvious and practical application of this technique is Supplier Assessment. This technique exploits the capability of parametric models to compare organization's normalized cost density or Manufacturing Complexity. If, on paper, you have very similar proposals, how can you judge which organization will provide the best value for money? Heritage data will determine the level of development that is required and Operating and Support Cost will establish the Through Life Cost, but if these cannot differentiate the proposals perhaps the productivity of the companies should be considered. This is not a

---

KCB and Mr Stan Porter, Question 29, <http://www.publications.parliament.uk/pa/cm200102/cmselect/cmpubacc/370/1111403.htm> (emphasis added).

simple comparison, but can be achieved easily on the basis of Manufacturing Complexity.

Figure 4.4 shows a simple graph comparing various items of similar technology. It clearly indicates which is the more competitive organization. This process of comparison will lead to suppliers knowing that since you are able to control them, they will need more control of themselves. This will enable the perfection of negotiation skills, leading to more success, as negotiation is based on accurate technology Cost Drivers. A parametric questionnaire (see Figure 4.5) is one tool to enable proposal validation in minutes. DEF FORM 143[3] is used by the UK MOD within software ITTs to gather parametric cost drivers data from suppliers which can be used to produce an independent estimate of the suppliers bid quickly.

All this information can be garnered at a very early stage in a project if required, providing better estimating accuracy for the purchased items. Normalization enables a bigger data sample as previous proposals received from the same companies can be compared. This in turn enables the establishing of trends. Are the suppliers becoming more or less efficient? Ultimately, these comparisons can be used to determine preferred suppliers.

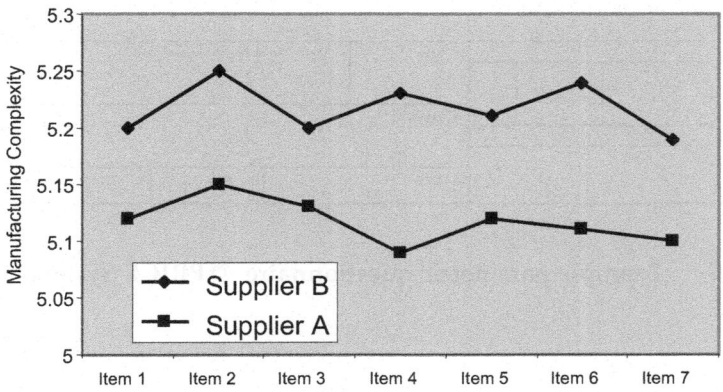

Figure 4.4 **Comparison of Suppliers A and B**

---

3   http://www.aof.mod.uk/aofcontent/tactical/toolkit/downloads/defforms/word/143.doc.

SOFTWARE PARAMETRIC (SYSTEM) INFORMATION

rev 2 Feb 96

1. CSCI/CSC NAME
2. SYSTEM NAME
3. WBS ELEMENT

4. OPERATING ENVIRONMENT (CHECK ONE) *(Keep in mind this describes not as much the location of software, but specifications it must meet.)*

- [ ] a. Production Center - Internally Developed
- [ ] b. Production Center - Contracted Software
- [ ] c. Military Ground
- [ ] d. Military Mobile
- [ ] e. Commercial Avionics
- [ ] f. Mil-Spec Avionics
- [ ] g. Unmanned Space
- [ ] h. Manned Space

5. INTEGRATION FACTORS

Internal Integration Factor- the level of difficulty of integrating the units to CSCs and the CSCs to the CSCI. Enter a decimal between 0 and 1.

External Integration Factor-the level of difficulty of integrating this CSCI with others in the system. Enter a decimal value between 0 and 1.

(See table below for typical examples.)

EXAMPLES OF FACTORS	LOW TEAM EXPERIENCE	NORMAL EXPERIENCED TEAM	HIGH EXPERIENCED TEAM
Minimum Coupling&Timing Constraints	0.5	0.3	0.2
Strict Coupling & Timing Constraints	0.7	0.5	0.3
Strictest Coupling&Timing Constraint	1	0.7	0.5

6. UTILIZATION - Fraction of available memory/cycle time required by software. Use only hardware constrains development effort by either memory or timing.
Enter fraction of total available capacity utilized. Values of 50% or less have no effect on cost. The upper limit is 95%.

a. Percentage of available speed utilized   b. Percentage of available memory space utilized

7. SCHEDULE *(Either SDR or SSR is mandatory. Other dates are optional.)*

- SCON The date the System Concept effort starts
- SRR The date System Requirements Analysis completes
- SDR The date System Design Review completes
- SSR The date Software Specification Review completes
- PDR The date Preliminary Design Review completes
- TRR The date Test Readiness Review completes
- FCA The date Functional Configuration Audit completes
- PCA The Physical Configuration Audit completes
- FQR The date Formal Qualification Review completes
- OTE The date Operational Test & Evaluation completes

8. COMPLEXITY *(Check one in each category - personnel, environment, and mnagement)*

a. PERSONNEL
- [ ] Relatively Inexperienced-Many New Hires
- [ ] Mixed Experience-Some New Hires
- [ ] Normal Crew-Experienced
- [ ] Extensive Experience-Some Top Talent
- [ ] Exceptional Crew-Best Talent

b. ENVIRONMENT
- [ ] Old Hat, Redo of Previous Project
- [ ] Familiar Type of Project
- [ ] Normal New Project

c. MANAGEMENT FACTORS
- [ ] Multinational Project
- [ ] More Than One Location/Organization

9. PREPARED BY LAST NAME FIRST MI
10. DATE
11. PHONE *(Include DSN or Area Code)*
12. TITLE
13. ADDRESS
14. CITY **15. STATE** **16. ZIP CODE (9-digit)**

Figure 4.5   Example parametric questionnaire. © PRICE Systems LLC

Summary

One of the main attractions of parametric estimating is the ability to estimate with speed without the loss of accuracy. This is particularly important during bidding and tendering for new work. In this chapter you will have observed that:

- Parametrics can be a great asset when making bid/no-bid decisions.

- Parametric models provide a cost estimate and an estimate of the optimum schedule for the activities.

- Schedule penalties can be applied if customers' demands are not compatible with the optimum schedule.

- It is possible to compare efficiencies of companies that produce similar technologies for supplier assessment and inclusion into your proposals.

- Parametric questionnaires are efficient means of gathering parametric information.

5

How to ... Prepare a Focused Business Plan

The Business Plan of an organization is a focal point for staff and management alike. Managers responsible for functions or departments can take their direction from the Business Plan to determine their individual Operational Plans. The Operational Plans will refer to the Business Plan, but will centre on the goals of the department, work group or function. Project Managers will refer to the Business Plan in their Project Plans. The Project Plan will have the individual project goals at its heart, but will reflect the ultimate needs of the Business Plan in terms of profitability, public relations or successful outcome of the project.

So with such a high profile, what is a Business Plan? And how can parametric estimating help in the formulation of a credible, achievable Business Plan?

Business Plan

A Business Plan is a statement, usually articulated in a formal document, which states the business goals of an organization. This is expanded into the reasons why these business goals are thought to be achievable and attainable in the period covered by the Business Plan, normally a three-year period. Finally, it will cover the plan for reaching those goals. Sundry information will cover background about the organization and the team involved in attempting to reach the stated goals.

The goals are not always financial: both profit and non-profit organizations will write Business Plans for the reasons of staff communication as stated above. Generally, not-for-profit organizations will focus on service levels to be achieved and profit organizations will focus on financial goals.

If an investor is needed to fund the Business Plan, then the document needs to be constructed with a clear proposal, convincing discussion and a strong conclusion. The plan will need to be supported with detailed facts and figures projected over the relevant period. The Business Plan will need to make a case for funding by demonstrating a good grasp of the financial matters, making realistic assumptions, considering all the relevant factors and stating that there is a good chance of successfully achieving the goals in the future.

Parametrics

For the Business Plan to motivate and inspire staff it needs to be realistic; unrealistic business goals will be seen as impossible. Parametric estimating can offer several applications which can help provide realism. These are briefly described here and detailed in the following sections:

1. *Winning productivity or not* – Having determined, in Chapter 2, that parametrics can be used to determine productivity, it is a natural progression to be curious as to which technologies an organization excels in and which it does not? The comparison of Manufacturing Complexity for hardware or Organizational Productivity for software against average industry will establish without doubt where investment needs to be made and when the organization has a competitive edge.

2. *Productivity tracking* – Productivity is not a static phenomenon; once the productivity of a technology in an organization has been established, it can be tracked. If a Business Plan is making a case for new investment, the return on the investment should be a marked improvement in productivity. However, this can be difficult to map when products or design change; parametrics will help.

3. *Competitor analysis* – To provide a realistic Business Plan there needs to be an element of innovation to excite the staff in the organization and investors. However, it also needs to be marketable. The ability to determine the cost of competitors' technology will help to frame the Business Plan in terms of where the technologies would fit in the market place. Businesses can fail, not through lack of ideas, but their inability to compete and differentiate themselves from the competition in terms of features and cost.

Productive or Not?

Comparisons with the productivity of an average industry will naturally indicate a technology which would be preferred in the market place. It also determines the competitiveness of the products placed in the market. The use of an average scale of productivity as the benchmark implies that some organizations will be above and some below the standard. Establishing where your organization lies and in which technologies you excel should influence your Business Plan.

For example, Figure 5.1 shows a series of Printed Circuit Boards (PCBs) for which their Manufacturing Complexity (normalized cost density) has been determined. One particular type of board is consistently above the average industry value expected for this type of technology. This indicates that the productivity of the organization for this type of technology is not as good as average industry. The average industry value is determined through calibration of a cross-section of manufacturers of this technology or Manufacturing Complexity calculators provided as utilities in parametric models.

The future promotion or marketing of this technology should be reconsidered until investments in manufacturing processes or tooling have yielded a better productivity. Trying to sell these particular PCBs will be fruitless as other organizations will have the productivity advantage and be able to undercut the price of the items offered.

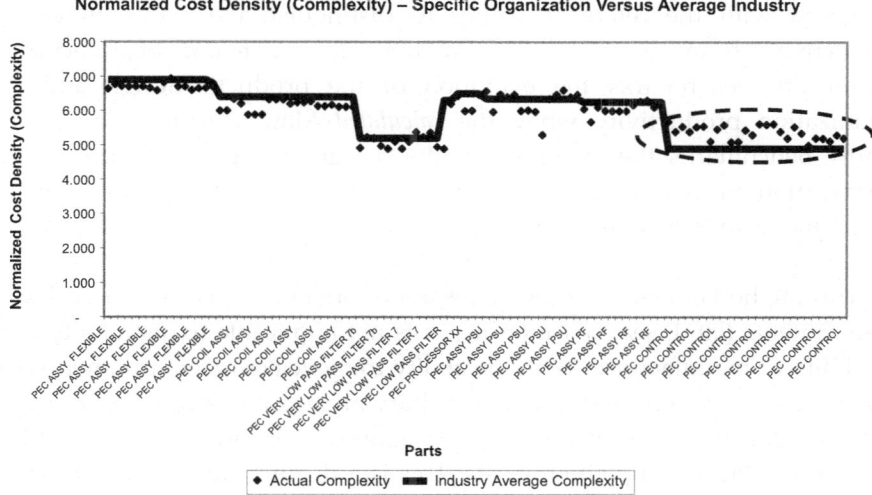

Figure 5.1   Productivity analysis compared to average industry

Productivity Tracking

Benchmarking the productivity of technologies as described above should not be a one-off activity. It is possible to monitor or track the productivity of an organization over a period of time. The benchmarking process should follow the development of an organization's productivity relative to the calculated Manufacturing Complexity based upon industry average figures. There are two ways to generate a Manufacturing Complexity: one, through effectively *reversing* the cost estimating model to establish the input parameter for the Manufacturing Complexity – the calibration process; two, by calculating the normalized cost densities through observing the design influences that alter the complexity as a result of design changes.

The first stage in this application is to define and construct a productivity complexity based on a family of stable products and the most representative of the organization's activities. This can be achieved by establishing a reference value for the productivity index using the weighted combination of the Manufacturing Complexities. At set periods this productivity index will be calculated from the cost to completion of all projects and comparing this with the reference value for the productivity of the organization. Analysis of the possible differences and research of causes (suppliers in monopoly situations, new organization of the work, requirement creep, modified schedule of production, and so on) will yield the explanation of this benchmark.

The updated productivity index (Manufacturing Complexity) can be compared with the reference values to distinguish the evolution of the productivity from the technical evolution. The *calibrated* Manufacturing Complexity benchmarks the evolution of the product's design and the organization productivity while the *calculated* Manufacturing Complexity simply benchmarks the evolution of the design taking no account of the organization productivity. The difference between the two represents the critical measure (see Figure 5.2).

This method of benchmarking a project or organization can be used for all environments (land, sea, space or air); it is not applicable to just one type of capability or manufacturer. With the ability to move across environments it is possible to compare commercial and military organizations equally. Decisions and judgements can be made between commercial and military suppliers where no military competition exists. This is a distinct advantage over other alternative approaches to monitoring productivity.

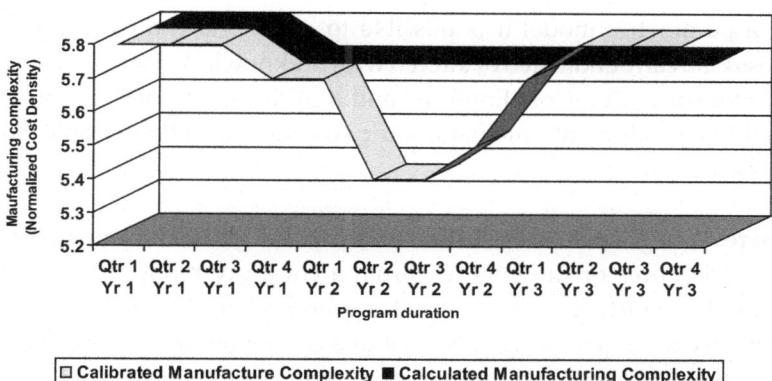

Figure 5.2 Productivity tracking

Furthermore it is possible to assess the productivity of an organization or country at system, sub-system and equipment levels. The technique can be used to compare the whole cost or just the labour cost if that is the focus of the efficiency drive.

This technique is not limited to hardware either; the approach is equally applicable to software; in which case the productivity index is the Organizational Productivity. The Organizational Productivity can be monitored and reviewed at the program level. Naturally a project with a combination of hardware and software technologies can be monitored using both measurements.

Using this technique to benchmark a third party rather than itself, an organization can determine preferred suppliers and monitor their change in productivity over the course of a contract. It is also possible for an organization to demonstrate its changing efficiencies when its customer is sceptical about the organization's productivity or its drive to improve productivity.

In the context of the Business Plan this parametric solution will provide an independent measurable target for the organization over time.

Competitor Analysis

Using a parametric model it is possible to conduct a competitor analysis. A Business Plan can be adjusted or altered with the knowledge of what competitors' products cost. With a Cost Engineer and a Systems Engineer to interpret the competitors' product into the parametric cost model, a competitors' likely cost can be estimated.

There is a great deal of information to be gleaned from trade shows, product brochures or the internet. From a cut-away model (Figure 5.3) on a trade stand or a picture on the internet, a good Systems Engineer can determine the likely technologies being used within a competitor's product. They will be able to break down the equipment or system into its technologies based on their experience of or by analogy to their own organization's products.

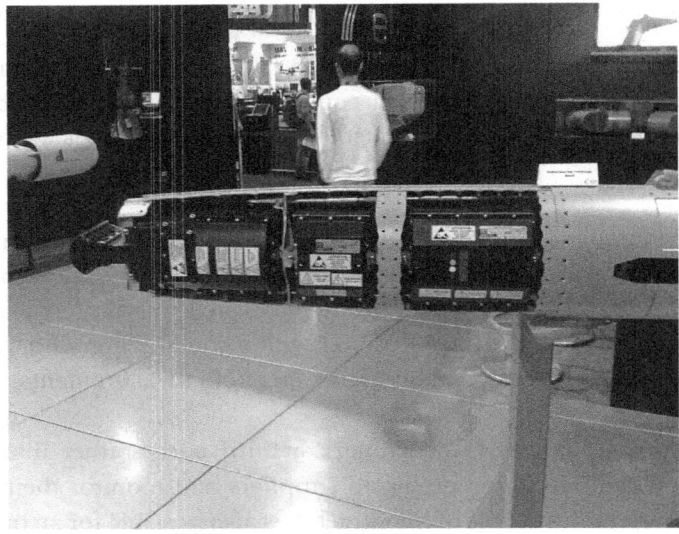

Figure 5.3   Product cut-away. © Dale Shermon, Singapore Airshow

Similarly, weights are often provided in technical documentation or specifications on competitor's websites. From this information a Systems Engineer need only provide an indication of how the top-level weight is distributed by technology so that the Cost Engineer can begin to create a Product Breakdown Structure (PBS) (see Figure 5.4) and populate the Cost Drivers to generate a competitor's cost.

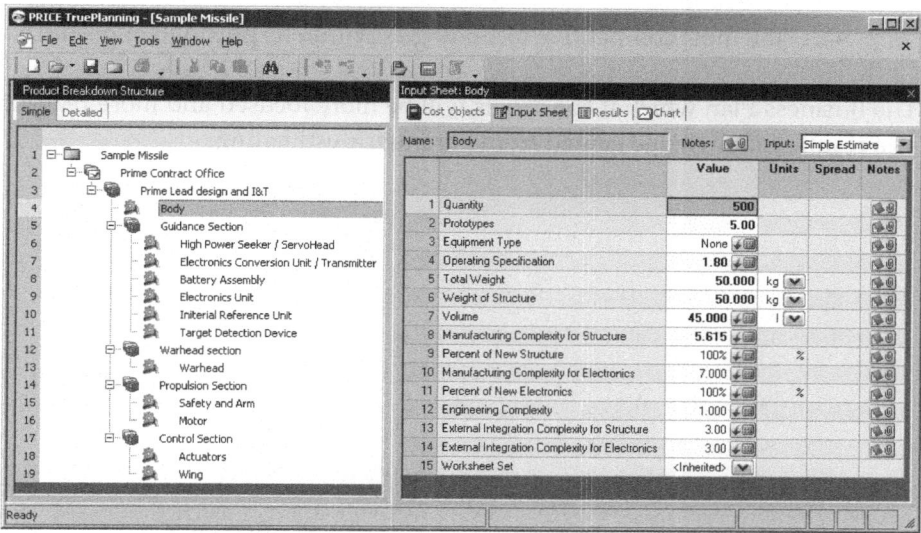

Figure 5.4 Missile PBS

Other assumptions, such as the production quantity, production start date, percentage new design, development team experience, economic conditions and so forth can be based on assumptions about the organization's own products, thus making the parametric cost estimate directly comparable to its own offering.

Assuming the competition has average productivity, or by conducting some calibration to establish their productivity relative to average industry and applying that factor, it should be possible to predict their costs with confidence.

Summary

It is possible for an organization to make strategic decisions for the future of its business with enhanced cost information regarding:

- competitors' costs;

- productivity of its programs;

- the most competitive technologies within its portfolio of products.

The Business Plan which it constructs will be more focused and more realistic as a result of knowledge gained from parametric estimating.

6

How to ... Validate Quotations from Suppliers

One of the most effective applications of parametric models is derived from the ability to produce normalized cost density or Manufacturing Complexity data. This significantly broadens the number of projects that can be used for comparisons when proposals are received from suppliers.

The fact that Manufacturing Complexity values are an empirical representation of the technology used in a product (along with the productivity of the manufacturing organization) is important. If the technology is held constant, then the only difference between two suppliers providing the same technology is the productivity of those organizations. The supplier with the lower Manufacturing Complexity will be the preferred supplier (see Figure 6.1).

One of the applications that will provide the quickest and most obvious return on investment is proposal validation. This involves the analysis of proposals relative to the historical complexities that have been paid for in the past – an application of parametrics called Supplier Assessment (see also Chapter 4), which helps procurement staff to make sure they will pay the right price for what they buy.

One of the main issues in proposal analysis is the comparison of non-consistent data. The solution is normalization, which means the extraction of what comprises the differences. In this example, the cost on the proposal is first normalized by converting into a Manufacturing Complexity and then compared with the reference range. This is a process that needs to be implemented for all the proposals. Proposals that fall inside the range show costs that are valid.

However, the calibrated Manufacturing Complexity can also fall outside the range; this can be due to three reasons linked to the consistency with the requirements and the design (see Figure 6.2). The first reason why a high Manufacturing Complexity may occur is because of a high level of design complexity resulting from the specification. The cost is justified and the element must be put in another family of technology. The second reason is high complexity due to over-design for which the solution is a re-design. The third reason is a spurious level of complexity, with no justifiable reason, and in this case the cost can be negotiated.

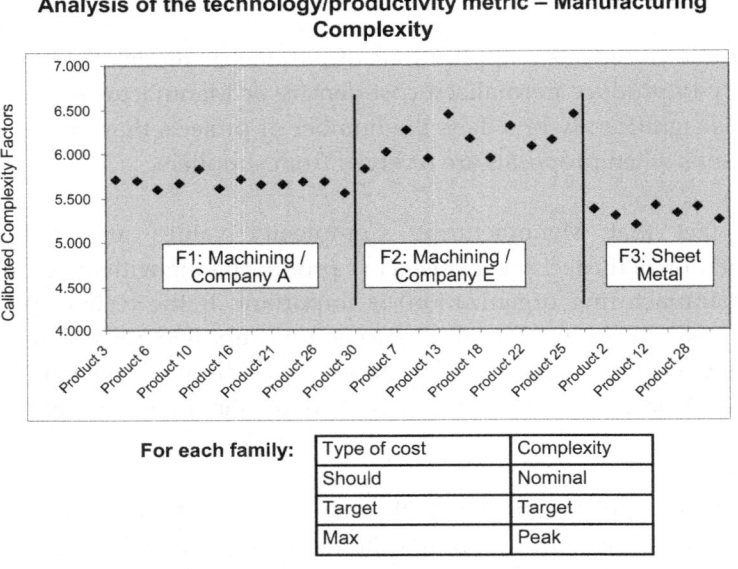

Figure 6.1   Establishing a complexity for a family of technologies

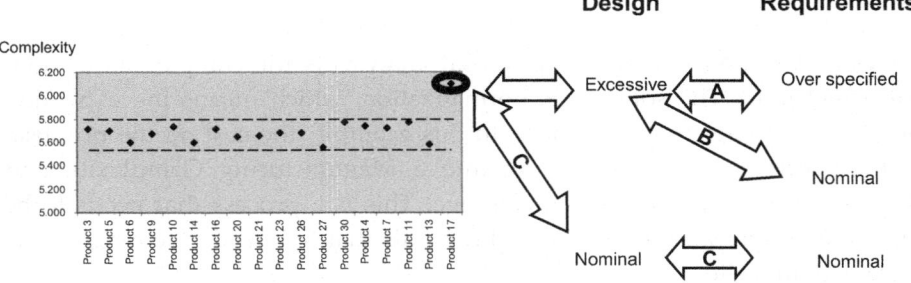

Figure 6.2   Looking for explanations in the design or requirements

Supplier Assessment

Parametric model vendors have developed a unique methodology to assess submissions for suppliers rapidly in order to determine if they are reasonable and provide best value for money. The parametric cost model offers great benefits in the procurement of equipment – both subcontracted and Off the Shelf (OTS) items. The model enables:

- the establishing of cost targets for Requests for Proposals (RFP);

- the evaluating of bids;

- the selecting of the most appropriate suppliers.

There are many factors that influence the cost to produce an item. To make meaningful cost comparisons between different items and suppliers will often require adjusting 'raw' costs in a 'normalization' process. A simple example of this is production batch quantity. A production run of 3,000 is likely to result in a significantly lower Unit Production Cost (UPC) than a run of 300, but a comparison of UPC fails to show whether the costs are *reasonable*. Alternatively, different sizes of the same technology prevent direct comparisons. Other elements such as exchange rate, production rate or technology improvement makes the analysis and comparison difficult. The Supplier Assessment solution is based on the use of the normalized representation of the cost, the complexity factor.

The evaluation process starts with an analysis of the current situation, via the calculation of the Manufacturing Complexity of the purchased items; this is called product calibration. The costs are translated into normalized values, using complexity.

Many parametric model users have made significant cost savings in their procurement activity using this method. The qualitative benefits of the process are:

- suppliers know that you are able to control them – they will need more control themselves;

- better estimating accuracy for purchased items at the concept phase;

- proposal validation in minutes, detailed evaluation if necessary;

- different negotiation material based on technology Cost Drivers;
- normalization enables bigger data samples across industry;
- quick return on investment.

PREFERRED SUPPLIER

How are preferred suppliers identified? Now it is obvious. The supplier who has the track record for the lowest complexity is the preferred supplier. This does not necessarily require the elimination of all the other suppliers from the organization's records: once the resources of the preferred supplier have been exhausted, the services of the next most productive supplier will be needed.

Many competitive proposals open with: 'As the world's leading provider of ...'. This type of unsubstantiated claim is not nearly as convincing as a detailed analysis of an organization's competitors which identifies the organization itself as the most productive and thus identifies it as a candidate for the role of preferred supplier.

Parametrics and the Supply Chain

Parametric models can be used at any level in the supply chain; whether by a government procurement agency spending millions of dollars or by a specialist sub-contractor providing particular equipment. The common approach is shown in Figure 6.3, where cash flows down the supply chain to acquire goods and services are integrated as they travel up the supply chain, hopefully with each layer adding some value. The information which travels both up and down the supply chain can sometimes be misleading, particularly when it comes to costs. A parametric cost model, adopted by a supply chain will provide a common language which allows information to be shared.

Government defence departments have taken the initiative to stipulate a parametric cost model which is now used for all major proposals. It enables them to focus on the inputs and outputs of the cost model, rather than the difference in the cost modelling between suppliers. It also ensures that the parametric cost model used to present the proposals to the customer is identical. This ensures that the supply chain has a common understanding of all the Work Breakdown Structure activities and the Organizational Breakdown Structure resources, but leaves the suppliers total freedom to propose any configurations of product or system using the Product Breakdown Structure. With this approach everyone, suppliers and customers, involved in the project is aware of the production cost, for example, and its definition. There are no questions such as: is the tooling cost included or not? Or has the development cost been amortized over the production program or not?

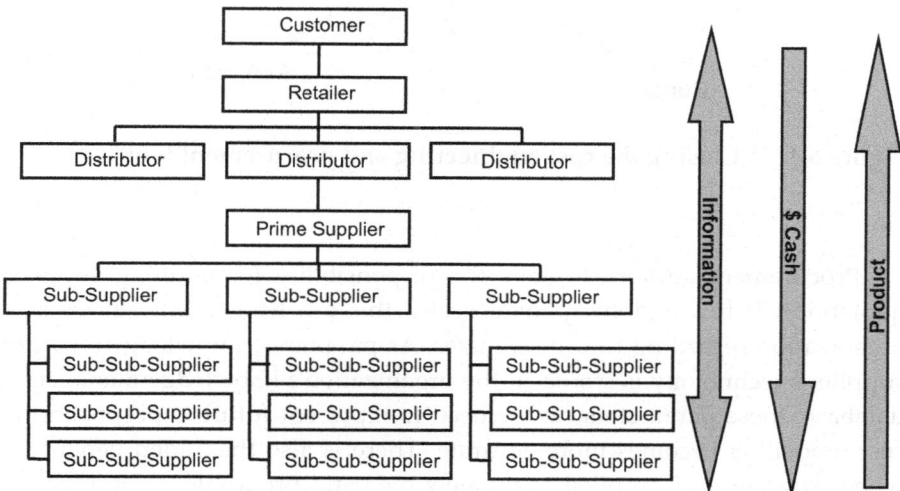

Figure 6.3   Example supply chain

The use of parametrics in the supply chain avoids commercial taboos. As soon as costs are discussed, the barriers come up between competitors. This makes the open discussion of different solutions to a customer's requirement very difficult. Using parametrics the actual costs can be substituted for parametric Cost Drivers which describe the technology of the hardware, complexity of the software and the magnitude of the hardware and software. This is an area which engineers feel more able to discuss with other elements of the supply chain without feeling commercially exposed.

Finally, the use of parametrics in the procurement function helps to refine the information available for the estimation of future bids and proposals. This application of parametric models in procurement closes the loop, as seen in Figure 6.4, and provides for more successful and accurate estimates.

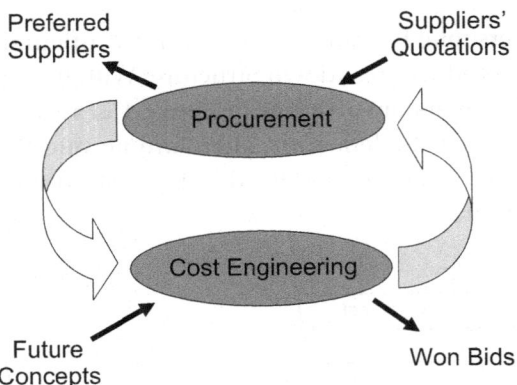

Figure 6.4 Closing the cost engineering and procurement loop

Procurement authorities have the responsibility for auditing suppliers' quotations. If this is done parametrically, this can lead to efficiencies in an organization regarding future estimates. As parametric knowledge regarding supplier's technology is stored in the organization's knowledge management database, these parameters can be used to represent future concepts and the cost modelling becomes more accurate. There is also the added benefit that when parametrics is applied to the new bids and proposals, a proportion of these will be successful and lead to the placement of new contracts. Without doubt these new contracts will become projects which will need components, equipment and systems to be acquired or purchased. If these bids and proposals have utilized the knowledge from previously investigated programs within the procurement function, then these new projects will hopefully already have been based upon the preferred supplier of the procurement function, in due course making their task easier.

Summary

If your suppliers are selected on an ad-hoc basis then you are less competitive yourself. In this chapter you have learnt that parametrics can help:

- Selecting preferred supplier can be conducted on a rational basis.

- Bids can be evaluated quickly.

- Realistic and achievable cost targets can be specified in your ITTs.

- Bids can be compared on a consistent basis.

- In manufacturing complexity terms the difference between two suppliers providing the same technology is the productivity.

- Parametric models can be used at any level of the supply chain.

- A stipulated parametric model can provide a common language between organizations and countries.

- Detailed discussion around parametrics can avoid commercial taboos.

- Usage in procurement or purchasing can improve accuracy of future bids.

7
How to ... Manage a Program Effectively

I drive to the office when necessary to attend meetings, teach classes or meet customers. This trip involves the M4 motorway and a series of A-class roads. The distance is 81 miles and will take 1 hour and 20 minutes. Why is this relevant to project cost control?

Cost control on a project is the discipline of reconciling planned cost or man-effort to actual cost or man-effort while considering the achievement of the project. This combination of planned, actual and achieved provides the basis of Earned Value Management (EVM).

Before we start to look closely at EVM and my journey to work, it is worth briefly mentioning other methods of cost monitoring and control. Cash flow management is the discipline of accountancy which ensures that liquid capital is available to the working of the organization. It is of little interest to the average employee that a project has accrued a healthy income. The average employee cannot buy their weekly groceries with accruals. Cash into an organization is its life blood; it enables the organization to survive. Many profitable companies have been dissolved due to lack of cash flow.

A consistent cash flow is a time-driven indicator that the project is running acceptably. It is also important that milestones are met to enable milestone payments to be claimed. These are event driven achievements and form the basis of a cost management process. These milestones need to be tangible and well defined; Cost Engineering can often identify solid achievements which would make good milestones.

The Baseline

It is critical to the successful and fruitful outcome of a project to determine the baseline correctly. The customer will normally provide a Statement of Work (SOW) which indicates the general direction of the project in its broadest form and desired outcome. The Work Breakdown Structure (WBS) created during the proposal will identify the activities that need to be successfully accomplished to enable a technical well-engineered solution to be the end result. These activities can be grouped into convenient Work Packages. They are normally assigned to technical leaders responsible for the work and may not correspond to a milestone. Therefore, it is the project manager's responsibility to manage the work in such a way that payments are received regularly and that technical progress is made.

The Work Packages containing the activities will have resources allocated to them. These resources are part of the Organization Breakdown Structure (OBS) of a company. They consist of the labour, material and other direct costs that will be consumed during the conduct of the project. The intersection of the WBS and the OBS is a Control Account (see Figure 7.1).

Establishing a cost baseline for this project will result from the proposal estimate. It will be necessary to take the winning proposal and possibly rearrange the cost estimate into the Control Accounts if the customer structure

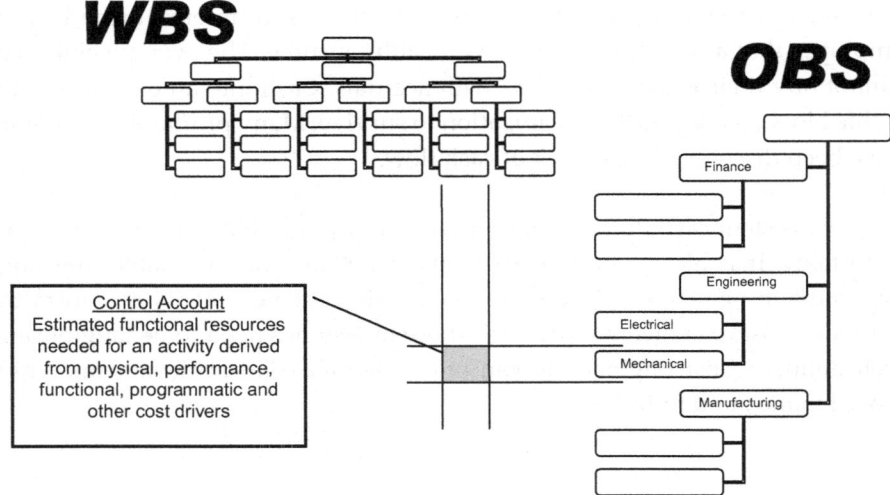

Figure 7.1 Definition of a Control Account

for the proposal was not aligned to the originator's OBS and WBS natural structure.

Earned Value Management (EVM)

Having determined the baseline resources for the Control Accounts, the project can begin. Tracking time is easy: days pass and weeks pass when difficult technical problems need to be solved. Accounts departments with timesheets ensure that costs are gathered and religiously allocated to the correct Control Accounts. But what about performance?

Figure 7.2 is generally recognized as the time, cost, performance triangle of projects. The three elements each have an influence upon the others. What is required is a technique to assess the performance of a project. Earned Value allows us to do that. It also enables organizations to forecast the cost and time to completion, but unfortunately not performance.

There have been many books written about the Earned Value Management (EVM) project control technique. It is not my intention to detail the theory here, but to consider the basic principles, then to explore the advantages of integrating EVM with parametrics.

Earned Value is also known as the Budgeted Cost of Work Performed (BCWP). The BCWP needs to be determined accurately to ensure that any performance indices that are calculated to enable future projections are accurate. It is therefore important at the outset of a project to spend considerable time and effort establishing credible BCWP values from the proposal that technical leaders are prepared to commit to. It can take weeks of effort to determine the

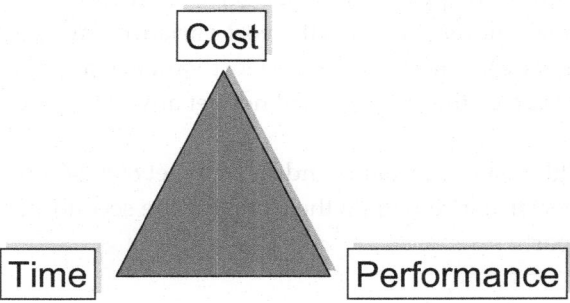

Figure 7.2   The time, cost, performance triangle

correct baseline, and maintaining this baseline can be very time consuming. In some cases, armies of staff have been employed on projects to maintain this baseline.

Once the project is underway the Actual Cost of Work Performed (ACWP) can be monitored. This represents the actual cost of completing the work itself in any given period of time. From these two pieces of information it is possible to determine cost variances which are the difference between the Earned Value and Actual Cost. The cost variance is the BCWP less the ACWP. If the outcome is positive, we have an efficient project and if it is negative then action needs to be taken.

For my journey to work the Budgeted Cost at Completion (BAC) is 1 hour and 20 minutes or 80 minutes. The Membury service station on the motorway marks the half way point to work. (It is next to the 500 foot Membury transmitting station for broadcasting and telecommunications, which I find hard to miss even on the earliest of mornings.) I know that my Budgeted Cost of Work Performed (BCWP) at this point needs to be 40 minutes.

If I manage this first part of the journey in 45 minutes, the journey variance would be the Earned Value (40) less the Actual Cost (45), resulting in 5 minutes – not good progress. Rather than a variance, EVM systems would also determine a performance index which is given as a ratio of the Earned Value divided by the Actual Cost, or 40/45, which is 0.88. A performance index of less than one is not good and a performance index of greater than one would indicate good performance. Either calculation is beginning to indicate that I might be late for work. A similar calculation can be conducted for the schedule of projects.

It is useful having the parameters and indices of past performance, as these can be used to estimate the project outcome assuming the same performance to the completion of the project. My Budgeted Cost at Completion (BAC) has not changed at 80 minutes; if I use the cost variance, my absolute variance is the best progress I can make, or BAC + Cost variance at present = 80 + 5 = 85 minutes. This assumes that matters will not get any worse for my journey.

If I utilize the cost performance index, I can get 80/0.88 = 90.9 minutes. This assumes that performance will be the same for the second part of the journey as for the first part.

So why do I set a satellite navigation system in my car every morning? I drive the same route to work and I have never missed the Junction 11 turning from the M4. The answer is that the journey is never the same each day. The satellite navigation system is integrated into the car and has been preset with the legal speed limit on the various roads. As a result, when I start my journey it confidently predicts that I will arrive at the other end in 1 hour and 20 minutes. Using a map and a calculator would enable me to calculate the same BAC.

However, compared to a BAC calculated from a map, the satellite BAC is constantly updated. Traffic information is fed to the unit by radio signal to enable the computer to provide alternative routes, thus avoiding accidents and queues. Hence, my Cost Performance Indicator (CPI) might be 0.88 halfway through my project to transport me to work, but if the satellite navigation systems have adjusted the BAC to take into consideration a diversion or queue, it would be a mistake to phone the boss to tell him I will be only 10 minutes late for work. If the BAC has been altered to 100 minutes, the CPI could get me fired.

This is all fun, but it has a serious point. The baseline for a program needs to be monitored continuously to ensure that the estimate at completion is taking due consideration of the changing environment in which the project is being conducted. The problem, as stated earlier, is that constant monitoring of the technical baseline takes time and effort. Therefore, many projects will neglect the baseline or BAC and will continue to calculate the Estimate at Completion (EAC) on false assumptions. This is done with good intentions to keep down the perceived overhead of periodically re-establishing the baseline for the project.

Parametric Estimating and EVM

The need for parametric estimating to be the source of the baseline project is as convincing as my use of the satellite navigation system each morning to travel to work. Parametric models provide the cost, schedule, performance triangle described above. Their Cost Drivers, which are used to generate the cost and schedule estimates, are indicative of the performance and provide an excellent Earned Value progress measurement system.

Firstly, parametrics will ensure that the baseline is constructed logically and consistently, without the prejudice that can arise just because a department needs more work or is currently under-utilized. The estimating technique requires fewer resources and is quicker than bottom-up estimating, thus ensuring that more resources are spent scrutinizing the project. This also applies to re-establishing the baseline of the project, which will require a simple update of the Cost Drivers from the technical changes to the project, rather than a pause in project progress while the team updates their estimates.

Once the project is underway, monitoring the technical solution against the solution that was originally proposed will provide early detection and assessment of problems leading from the technical solution. This early visibility, like the SatNav will enable projects to steer towards success.

Parametric estimating engages many disciplines from an organization including engineering, accounting, schedule control, manufacturing and others. This naturally leads to an improved decision-making process from a holistic viewpoint, rather than the disciplines only considering the best solution for themselves. An integrated situation throughout the program, as shown in Figure 7.3, will lead to a better Estimate at Completion (EAC), using technical inputs rather than cost variances determined from a baseline which is constantly changing.

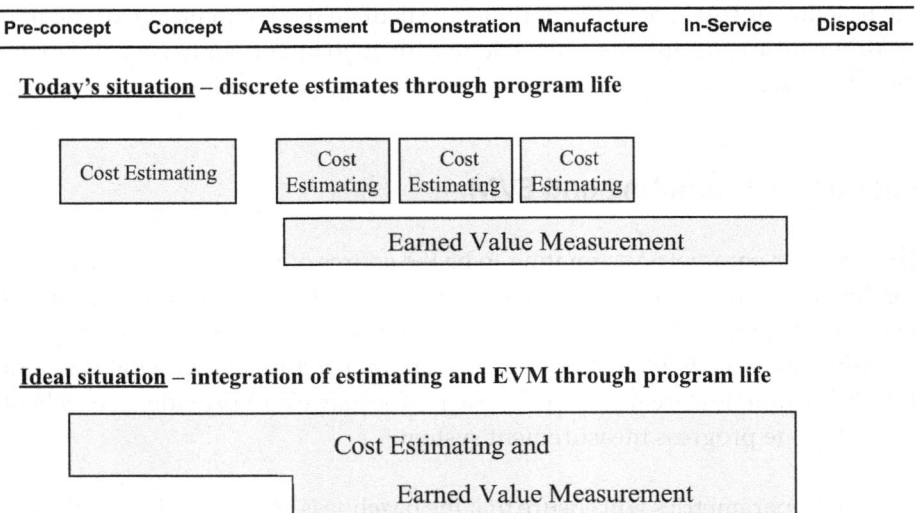

Figure 7.3   Integrated EVM and estimating system

Predictive EVM

PEVM is Predictive Earned Value Management – the 'marriage' of EVM project control with parametric estimating. By monitoring the technical changes to the baseline configuration the parametric model will create cost and schedule estimates continuously. This progressive scrutinizing of the technical solution which is being generated will have a direct bearing on the resulting cost and schedule of the baseline of the project.

Third generation parametric cost models are based on activity-based estimating methodology, which naturally lends itself to this project control technique. These parametric cost models have the ability to estimate to activity level and the resources required within those activities. In essence, the parametric model will estimate to the Control Account, a fundamental need of the EVM philosophy.

Using calibration facilities within the parametric models it is possible to monitor the health of the project in terms of its productivity. The Earned Value of the project can be calibrated in the parametric model and the productivity of the project established at that point of the project. This ensures that the EVM performance measurement baseline is synchronized with the actual achievement of the project to date. This then leads to the utilization of non-cost to cost relationships in the Estimate at Completion (EAC), which can be predicted using past performance data. The technical performance and measurements of the project are integrated, in PEVM, with cost and schedule through functional and physical parameter descriptions modelled in a parametric model.

Finally, the result of the combination of parametrics and project control through PEVM is more than the sum of two powerful techniques in themselves. PEVM adds analytical features that are unique, but proven as reliable, with the combination of linking performance, cost and schedule. This enables quick assessment for the Project Manager of cause and effect for their project, enabling real-time access to 'what if?' style question-solving.

PARAMETRICS AND EVM TOOLS

The interface between parametric estimating and EVM tools has already been demonstrated in a real-life example. The solution handled large-scale capital planning and management issues. The customer involved clearly understood that this solution would have a fundamental and far-reaching impact on the way they do their business. During the task, the program planning and control/EVM function was explored, together with how the solution could collect and report both contractor and 'organic' EVM data within a common framework.

Figure 7.4 provides a clear indication of the PEVM data flow. The flow of data through an integrated system reduces the resources needed to manipulate the data. It also ensures that there are no typographical errors.

The unique part of this integration is steps 7, 8 and 9. While it is normal for the EVM systems to have external data automatically fed into them from the accounts systems, it is unusual for the estimate to be calibrated from the EVM systems. An Enterprise Version of a third generation parametric model enables users to link seamlessly the power of parametric tools to other software tools using web services. This linkage connects the powerful forecasting tools of parametrics to widely used software applications.

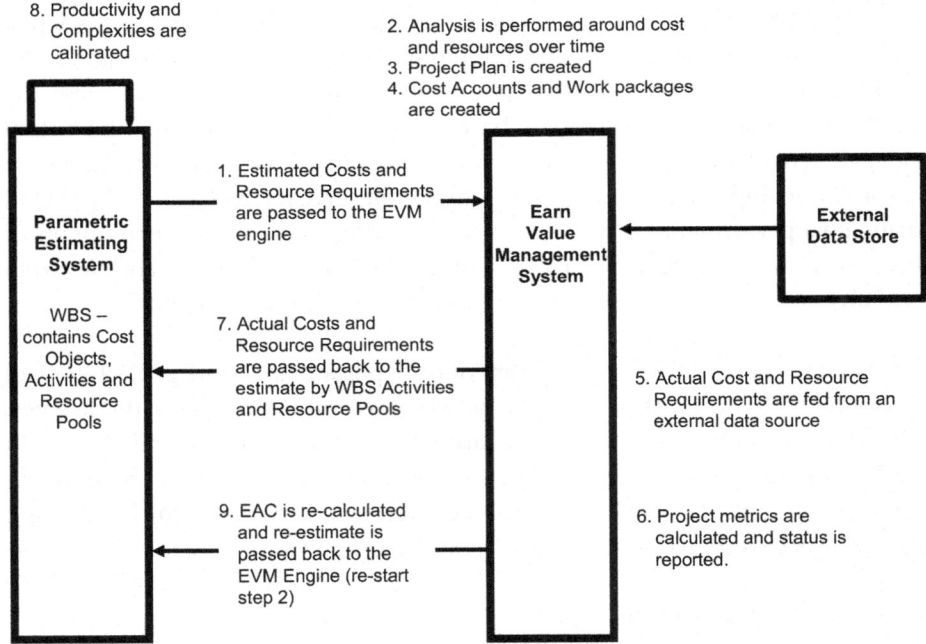

Figure 7.4   Predictive EVM integration

EVM SOFTWARE CASE STUDY

This case study will consider the application of parametrics to an EVM problem. The case study starts with the development of a simple COBOL product developed in the early 1990s. This is a simple Management Information System (MIS) – it contains a single Computer Software Configuration Item (CSCI) with 183,887 lines of code and an Organizational Productivity of 1.462 (somewhat more productive than the industry standard).

At the beginning of the project the baseline description is estimated using a parametric model and the budgetary estimate (baseline) of about $10.1m is set (Figure 7.5). The estimate, being based on activities and resources, will be used as the basis of the Cost Account.

As the program begins work, actual labour is recorded together with the progress. After the first seven months the program was expected to have spent $1.0m; but in reality the budget is exceeded by approximately $250k.

As a result it is possible to reflect this productivity by calibrating on the baseline model through determining the productivity factor that would yield a figure of $1.0m for the first seven months. The result is a productivity of 1.1, versus the 1.462 used to produce the budget. The conclusion is that we are not as productive with this project as we thought we would be (about 25 per cent less productive).

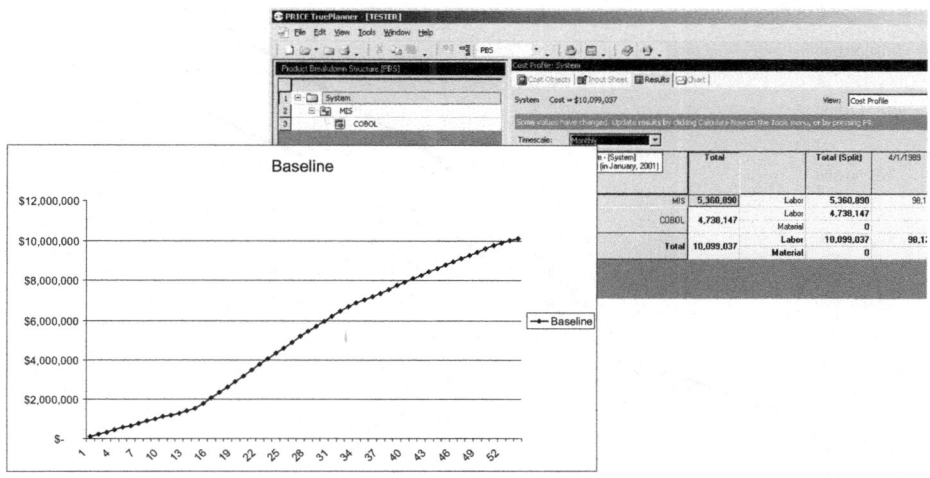

Figure 7.5   **Initial baseline estimate**

Following the productivity review at seven months, the recalibrated Organizational Productivity is applied to the baseline parametric model to estimate the cost to complete the project; this emulates what traditional EVM tools would produce since the productivity calibration to seven months Actual Costs serves as the Earned Value gauge – a lower than original productivity, as we have here, means less value earned than budgeted.

The Estimate to Completion (ETC) is approximately $12.3m, hence the Estimate at Completion (EAC), including the sunk £1.0m cost, will be a total project cost of $13.3m (see Figure 7.6).

When linking a parametric cost and schedule estimating model with an EVM tool the result is different. The Predictive EVM approach at the seven-month review would determine a different analysis. There are three initial project conditions that have changed after the first seven months of the project. Firstly, the target computer has changed from one that has been in use for over a year to one that, while fairly well known, is actually new to the developers. Secondly, a better understanding of requirements has changed the initial estimate of new size (Source Lines of Code) needed from 183,000 to 300,000. Finally, a clearer definition of interfaces has identified the need to include a COTS software package of 50K Source Lines of Code (SLOC) in the system.

These changes are captured as changes to the baseline software model. Again, calibrating the productivity factor to the $1m incurred to date, after

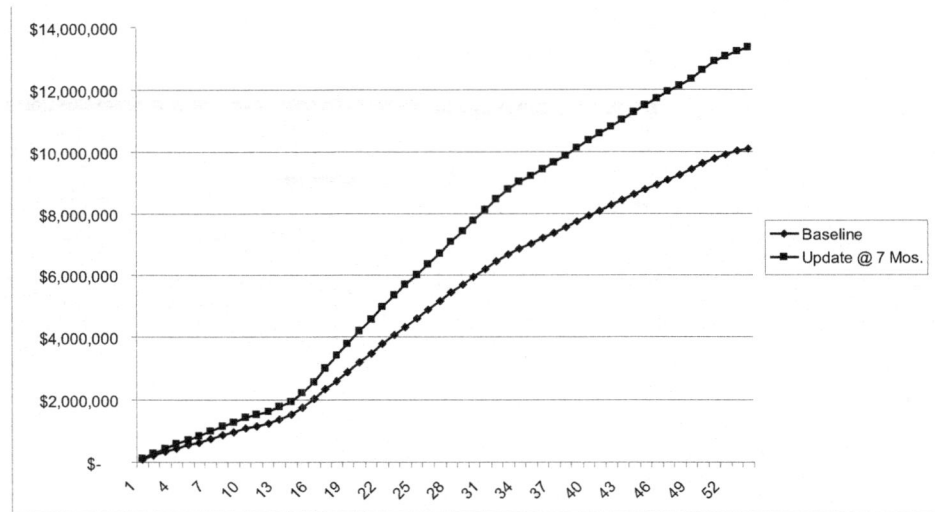

Figure 7.6   Progress at seven months

making the technical parameter changes to match the changed conditions, results in a higher realized productivity (1.58) than that used to produce the original project budget. Therefore, the conclusions at this stage of the project are:

- the project is larger than thought at the start;

- the project is over budget because it is larger than planned;

- the developers are actually performing more productively than planned.

With a Predictive EVM analysis the results of the seven-month review would be different, because of what has been learnt. The higher productivity is applied to the changed baseline in the parametric model to estimate the cost to complete the project; this predicts the Earned Value at project end on the basis of effort to date after it is aligned to the project conditions known at month seven.

In this case the ETC is approximately $15.5m, or a $16.5m total project cost (see Figure 7.7).

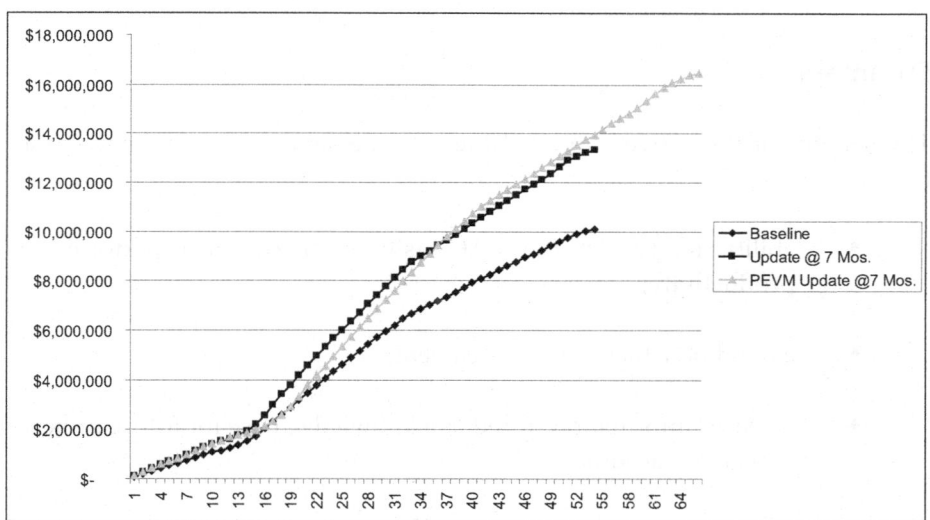

Figure 7.7 Predictive EVM prediction

In addition to the cost estimate the parametric model has a predicted schedule estimate. The changes to the productivity and technical parameters will add 12 months to the predicted baseline; it predicts that the project will overrun its schedule by approximately one year.

In this case study the estimates were produced with a parametric software model, but this is applicable to all aspects of systems, hardware, software, service and IT estimates. Calibration was accomplished by iterating on the model productivity factor until the costs for the first seven months were $1m, the Actual Cost for the first seven months.

Conclusions drawn from the PEVM analysis would resonate with most software project professionals. That is, that the project failure is the result of underestimating the size and scope of the project leading to significant schedule slippage being a more likely reason than the team being less productive than initially planned.

PEVM is the incorporation of updates to cost model drivers into traditional EVM to predict the cost at completion of a project. Because it is based upon the most current information about the project, as well as effort spent on it to date, it will produce the most credible prediction possible.

Summary

The benefits of Predictive Earned Value Management can be summarized as follows:

- ability to predict the EAC using non-cost past performance productivity;

- a baseline which is always current;

- projects' information stored in a knowledge base for future projects to increase accuracy;

- projects' success based on both the most current technical information and the current productivity of the organization;

- Engineering Change Proposal evaluation is an extension of normal supplier assessment analysis;

- enhanced variance analysis diagnostic capability – impacts related to non-Cost Drivers, size and complexity by deliverable end-items, Organization Productivity analysis;

- improved Independent Baseline Review (IBR) results – better risk assessments, better confidence levels;

- improved communication – cost and schedule performance actively linked to requirements and technical performance;

- ability to derive quantified contractor productivity;

- better ability to assess cost impact of schedule variances.

8

How to ... Achieve Accuracy in Cost Engineering

What is Accuracy?

Taking an assembly built up of equipment, it is possible to calculate the accuracy of a cost model in three different ways. These can be defined as follows:

- *Overall Tolerance* (at assembly level or on total of a series) – desired accuracy +/-5 per cent;

- *Average Tolerance* (the average of the absolute differences) – desired accuracy +/-10 per cent;

- *Detailed Tolerance* (maximum difference) – desired accuracy +/-30 per cent.

Accuracy is the tolerance of the possible output relative to the actual outcome. Calibration increases the accuracy of the parametric model, because it ensures that the model is more directly related to the productivity of the organization rather than any generic, industry average.

Accuracy of Results

Once a parametric model has been calibrated, the accuracy of the results are good. If a family of similar technologies are calibrated and an average complexity value determined for that particular technology, it is possible to demonstrate that the accuracy of the future estimate of that technology is within +/-10 per cent for a whole family.

Consider the example of a historical calibration which is based on five single and five triple voltage power supplies (see Table 8.1). The difference (UPC versus AUCOST) between the sum of the total costs is +2.3 per cent for single, -5.8 per cent for triple.

Table 8.1   Example of calibrated accuracy

Title	QTY	WT	MCPLXS	UPC	TXT1	AUCOST	Differ.	LEVEL 1 UPC	LEVEL 1 ATCOST	LEVEL 1 Differ.	LEVEL 2 Differ.	LEVEL 3 Differ.
AL841B	1	0.72	2.856	221	Single	249	-11%	221	249		11.3%	
AL896A	2	2.71	2.856	631	Single	499	26%	1 261	998		26.4%	
AL912A	2	0.8	2.856	235	Single	249	-6%	470	498		5.6%	
AL931A	2	0.975	2.856	276	Single	299	-8%	552	598		7.7%	
309-4907	2	9.5	2.856	1 744	Single	1758	-1%	3 488	3516		0.8%	
								5 992	5859	2.3%	9.7%	26.4%
ABTTS 1215	1	3.7	3.08	1 703	Triple	860	98%	1 703	860		98.0%	
EL 302T	1	7.5	3.08	2 817	Triple	2990	-6%	2 817	2990		5.8%	
6303 AS	1	8.9	3.08	3 182	Triple	4250	-25%	3 182	4250		25.1%	
6303 DS	1	12.6	3.08	4 077	Triple	3480	17%	4 077	3480		17.2%	
AX 323	1	9.5	3.08	3 334	Triple	4459	-25%	3 334	4459		25.2%	
								15 112	16039	5.8%	34.3%	98.0%

The average of the individual percentage cost differences is 9.7 per cent for single voltage, but 34.3 per cent for triple voltage. However, the maximum difference is 26.4 per cent for single voltage and 98.0 per cent for triple voltage. This demonstrates a common phenomena in Cost Engineering: individual estimates may be high or low, but when combined the overall effect will cancel out these anomalies. It is recommended that systems are broken down into their constituent units to increase the accuracy of parametric estimates, as with any other estimating technique. Compounding errors will ensure the accuracy is increased due to mathematical averages. This means that if an estimate of one item happens to be +15 per cent of the actual cost, the estimate must be +15 per cent in error. However, if that one item is broken down into two estimating components, one component might be in error by +15 per cent, but another might be in error by -15 per cent, effectively balancing the errors. The more the item is broken down, the more likely that any errors will balance off against one another.

The second lesson learnt is that one of the items has been incorrectly categorised in Table 8.1. Product ABTTS 1215 is a single voltage instrument which is why it is underestimated and has adversely affected the overall accuracy. Accuracy of the estimate is dependent on the accuracy of the information used to produce it.

Calibration

The parametric model will produce an average industry cost and schedule, when used without calibration. This average may be acceptable to procurement agencies that are not prepared to specify the organization who will be awarded the contract, but wish to set a budget. It also provides an acceptable benchmark in industry. However, for a proposal, committing bid or an accurate estimate related to a specific organization, average industry values are not acceptable. Calibration is required to increase the accuracy of the cost model. This can take two forms: product calibration and organization calibration.

Calibration is time consuming and the benefits need to be clear before proceeding. Calibration is the first return on investment when a parametric model is acquired. This investment includes the cost of the research into the parametric model and time to calibrate historical programs. Approached correctly, product calibration offers a considerable source of valuable information about an organization. The organization's productivity relative to average industry can be determined. By comparing the normalized cost density of the organization against calculator values it is possible to establish its relative productivity. The calculators for Manufacturing Complexity, as described in Chapter 2, produce the average industry figure. This is the result of cost research based on a database of identical technologies made by different organizations, some which are highly productive at producing the technology and some, naturally, less so. The calculator produces the average figure, representing the technology produced by an average productivity organization.

Organization calibration offers insight into the distribution of the organization's costs. For example, this may expose any excessive expenditure in the area of project management relative to engineering or development relative to production.

Product Calibration

Product calibration is about creating a rule of thumb for the productivity of the organization's manufacturing capability. Running the model in reverse, entering the historical cost and technical parameters of past projects, enables the organization to determine the typical normalized cost density of the product produced (see Figure 8.1).

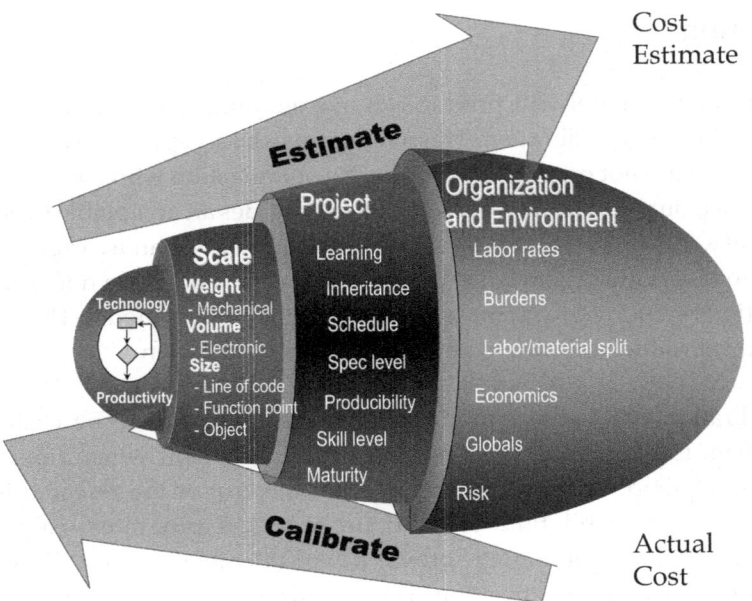

Figure 8.1 Calibration

Product calibration is a process of normalization. All the parameters of the model – organization, environment, project and scale – have an influence on the historical actual cost until the only residual is the technology of the hardware (know as complexity) or Organizational Productivity of the software.

CALIBRATION DATA GATHERING

The simplest way to gather information for calibration is to use a parametric questionnaire for internal or external data gathering. Having collected technical and cost data, the model facilitates the calibration process.

The calibration data can then be entered into a spreadsheet which makes manipulation relatively easy. Choosing to use graphical presentations, such as Figure 8.2, to determine any families of similar technologies makes them available for use in any future analysis.

Data analysis can transform a graph like that in Figure 8.2 into a significant trend. The first step in analysing the data is to consider the technology attribute. The Manufacturing Complexity is a technology index, therefore

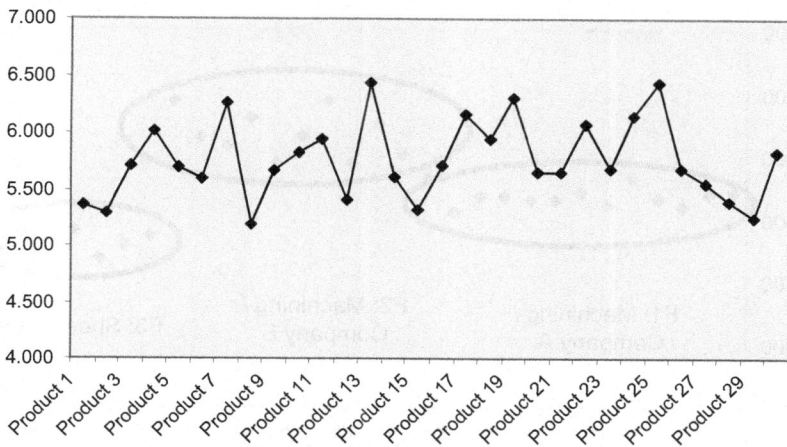

Figure 8.2   Results of calibration

grouping technologies into machined, sheet metal, casting, forging, composite, laminate, cables, Printed Circuit Board (PCB) and so forth will result in similar Manufacturing Complexity figures.

Secondly, these groups of technologies can be subdivided into other performance attributes. For example, fibre optic cables will occupy a group with a similar Manufacturing Complexity, but looking further into the number of fibres within the cable, reveals that the more fibres, the higher the Manufacturing Complexity. This is true for many technologies, motors are influenced by torque, engines by thrust or horse power, manufacturing process by the tolerance.

Figure 8.3 shows the transformation of the calibrated data once the technology families have been identified. In this case, there are three families: F1, F2 and F3. A subdivision of the organizations which produced technology; statistical analysis of the calibrated Complexity Factors, product clasification per homogenous families. With only one type of technology in the family, if there is any deviation remaining across the data set, this can be attributed to the producer. The technology could all be machined aluminium, but when the producers of that technology have been identified, some will be capable of producing the technology more efficiently than others. These families may be defined on the basis of product, technology or process type.

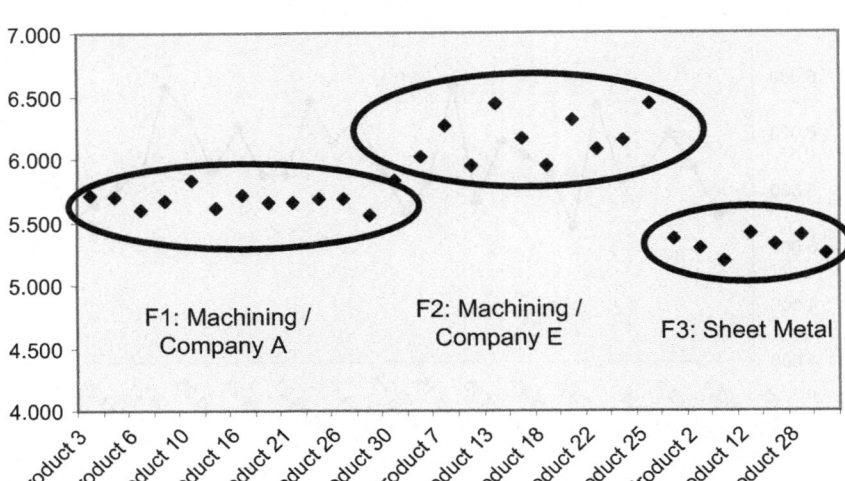

Figure 8.3   Identifying technology families

The average calibrated values can be used to estimate accurately the average industry cost of a particular technology by an organization. The coefficient of variation is an indication of the likely accuracy associated with the average value (see Table 8.2).

Table 8.2   Average complexity values for families of technologies

Family	Average	Coefficient of Variation
F1	5.671	1.4%
F2	6.170	2.9%
F3	5.315	1.5%

EVALUATION RANGE

Establishing a reference range for different future applications is useful, for example, for proposal validation, Cost as An Independent Variable (CAIV) as detailed in Chapter 16, and so on. Table 8.3 provides an example of the reference range including the nominal, target and peak Manufacturing

Complexities. A single point will be interesting for benchmark activities, but as we will discover, the possible range of Manufacturing Complexities resulting from a calibration exercise will put that point value into context. The range will determine the upper and lower boundary of what is normal for the Manufacturing Complexity. The use and derivation of the figures are as follows:

- *Nominal range* – this represents the average of the manufacturing complexities; when used for estimating it will produce the 'should cost' as it represents the average for the sample of data.

- *Target range* – this is determined by taking the lowest recorded calibration item or more reasonably the *nominal less one standard deviation*. The target range can be used to determine the target cost for Cost as An Independent Variable (CAIV) as detailed in Chapter 16 or Request for Proposal (RFP) activities. In both cases a challenging, but realistic, target cost is being sought to ensure that the in-house designers (in the case of CAIV) and the supplier (in the case of the RFP) are demanding but not demoralized by unrealistic cost expectations. This is not easy to achieve.

- *Peak range* – the peak range is a trigger mechanism for negotiation or justification of a supplier, if the source of the calibrated data is from a vendor, or the manufacturing department if the source of the comparison is in-house. The peak value represents the highest possible result from calibration of historical technologies; to exceed this value for a similar technology would indicate the presence of excess profit, monopoly or a simple misunderstanding of the pricing.

Table 8.3   Reference range per family of technology

Family	Nominal	Target	Peak
F1	5.671	5.59	5.75
F2	6.170	5.99	6.35
F3	5.315	5.23	5.40

GENERATION OF CALCULATED COMPLEXITY FACTORS

Parametric model vendors have large databases of historical project data. When this data is calibrated the result is the Manufacturing Complexity. This leaves the vendor with a dilemma: they can leave this Manufacturing Complexity visible to the user or mask it by conducting further analysis and presenting the output of the analysis to the user. This further analysis would look at the causes of the different manufacturing technologies. For example, in electronics the difference between component types (discrete, low-scale integrated circuits through to very large-scale integrated circuits) and circuit configuration (digital, analogue through to power) would cause a lower or higher Manufacturing Complexity. Consider Figure 8.4, where 50 structural items have been calibrated by a parametric model and their Manufacturing Complexities graphed.

At the outset, this looks like a scatter plot with no logic, all the items were produced using a similar manufacturing process, machined plate or bar stock using heavy milling. However, on inspection and analysis of the structural items it is possible to group those items which have a high precision, say +/-0.05mm or better, while items of lower precision fall lower down the scale.

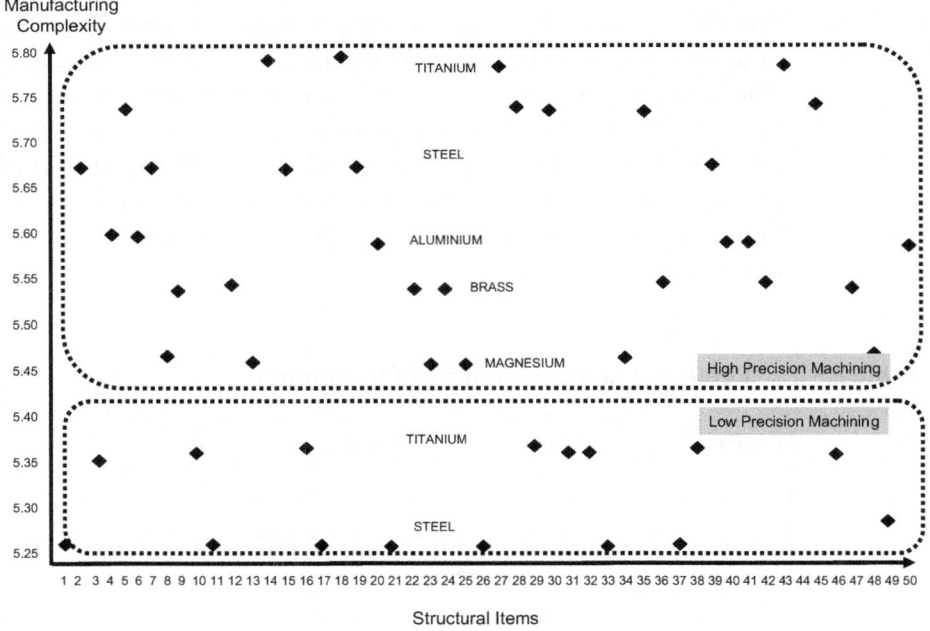

Figure 8.4   **Fifty calibrated structural items**

Further inspection reveals further grouping when considering the raw material for these items. At the top of the groups are the titanium and steel components, while at the bottom of the groups are the brass or magnesium components.

From these and more detailed observations it is possible to introduce a calculator that will predict the Manufacturing Complexity of technologies for future programs. For example, even with this basic analysis it is possible to see two questions which might arise in a simple structure complexity calculator: what is the precision of the future component and what material is it made from? If the calculator was given the answer 'high precision and aluminium', it could respond with the answer 5.55 to 5.60.

Using the detail calculator, it is possible to calculate Manufacturing Complexity based on cost research into the performance and design characteristics of the technology. This result can then be compared to the calibrated complexity and a *calibration factor* determined for specific technologies; hence the generic complexity calculator can be converted to an organization-specific calculator. Analysis of the variation between the calculated and calibrated complexity will result in the calibration factor per family (see Table 8.4) and increased accuracy for an organization.

This calibration factor reflects the productivity of the organization relative to the industry average. Using it, a benchmark may be produced for one

Table 8.4   Comparison of calibrated and calculated complexities

Product	Calibrated Manufacturing Complexity	Calculated Manufacturing Complexity	Calibration factor
Product 1	5.568	5.189	1.237
Product 2	5.284	5.084	1.273
Product 3	5.713	5.391	1.436
Product 4	6.014	5.792	1.265
Product 5	5.701	5.543	1.193
Product 6	5.596	5.391	1.263
Product 7	6.265	6.253	1.012
Product 8	5.182	5.084	1.127
Product 9	5.671	5.792	0.876
Product 10	5.831	5.792	1.043
Product 11	5.936	5.792	1.167
Product 12	5.402	5.084	1.461

technology of which the organization has intimate knowledge, which can then be applied to new technologies for which it has no history or experience.

ESTIMATION

From descriptive technical data of the selected products and the calibration factors (obtained from the calibration process) the product cost can be estimated parametrically.

ANALYSIS OF THE ESTIMATION ACCURACY

The variation between estimated and actual costs can be determined and analysed in this way. Analysis at element and system level of the breakdown structure will yield different accuracy results for different families of technology as described by the three levels at the beginning of this chapter (see Table 8.5). If these results are compared with the needs of the organization in terms of accuracy of its cost models, an implementation policy for the organization will be generated.

Table 8.5   Analysis of accuracy relative to accuracy definitions

Estimating accuracy			
Family	Technology 1	Technology 2	Technology 3
Tolerance level 1	3%	5%	2%
Tolerance level 2	9%	12%	7%
Tolerance level 3	18%	23%	16%

This completes the product calibration phase of the calibration of a parametric model. The results provide a set of parameters which reflect the productivity of the organization in total and the calibration process is equally applicable for hardware or software. The next phase will consider the distribution of costs across individual activities. This will reflect the individual differences between different organizations, hence the need for organization specific calibration.

Organizational Calibration

No two organizations have the same departmental or accounting structures. Therefore, the distribution of costs as specified in the parametric model will be specific to each organization.

During the development of the models it is important to have a common understanding of the historical costs. This involves determining the boundary between activities and resources to ensure the normalization process is accurate. This is used to define the categories of output cost in the estimates. These categories need to be tailored to ensure that they match the structure of the individual organization in question. This does not mean that the uncalibrated cost model is wrong (or that the organization is wrong), just that they are different.

Organization calibration opens the potential for further detailed investigations and tuning for accuracy. Consider Figure 8.5. It is clear, relative to the parametric model, that this organization might tend to spend too little in development prior to going into production, resulting in higher production costs.

Using organizational calibration across a number of historical programs it is possible to determine ratios that will modify the standard results from the parametric model to ensure that they align with the actual costs experienced by an organization. A number of sets of ratios might be stored in the parametric model reflecting the different sites within an organization, perhaps one biased towards development and research, while another site within the same organization might be biased towards manufacturing or mass production.

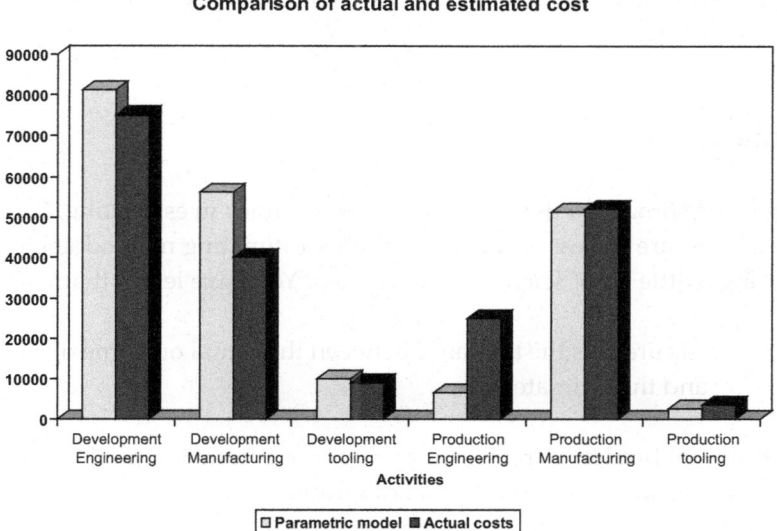

Figure 8.5 Organizational calibration

Experience has shown that there is likely to be differences between these ratios of different past programs; even when sites have been assessed, it is important to find a consensus for future application when estimating new programs. When the future programs have produced an estimate, it is important to ensure that the parametric tool is capable of estimating a range of outcomes as opposed to a single point estimate or deterministic estimate: that is, the program estimate together with a tolerance of accuracy, for example +/-20 per cent, to reflect the uncertainty. Producing a deterministic estimate can only guarantee one result – it will be wrong. And a point estimate cannot be right in the long term; it can only be proved to be incorrect.

Later in this book we will discuss risk analysis, but it is important to mention it here as the accuracy of the cost model can be increased by avoiding a single deterministic estimate in favour of a probability distribution, one that presents to the decision-maker a range of possible outcomes of the estimate and leaves them to establish how confident they wish to be in the certainty of the outcome.

The risk analysis feature of parametric models enables senior managers to establish their budgets, whether project or organization budgets, by defining the uncertainties that attach to them. The models can be used to perform Monte-Carlo simulations to generate a range of outcomes expressed by percentage probability. This feature enables users to establish the probability of exceeding an allocated budget, as well as running the simulations to establish whether the project is affordable.

Summary

Chapter 8 has broached the tricky subject of accuracy in estimating. Parametric cost estimates are no less accurate than other estimating methodologies, but it does bring a little bit of science to the process. You have learnt that:

- Accuracy is the tolerance between the actual outcome of a program and the estimate.

- Calibration increases the accuracy of parametric models; this can be product or organizational calibration.

- Product calibration will increase your understanding of your organization's productivity and technical drivers of the technology index through calculators.

- Organizational calibration will increase your knowledge of your cost distribution, providing a thumb-print of the organization or accounting structure.

- The simplest way to gather calibration data is with a questionnaire.

- Breaking an estimate down into a more detailed PBS can increase the likely accuracy of an estimate.

9

How to ... Accomplish Quality Assurance

While accuracy can be demonstrated using cost research techniques, this is not achieved in practice unless there is quality assurance to maintain discipline in the application of the models.

Multiple Estimating Methods

Cost Engineers are by their very nature sceptical. They are like accountants and use prudence when necessary. They will inevitably add costs to their estimates and look for senior staff to decide to remove those costs. Engineers are optimists and persuade themselves that everything is easy. If this was not the case they would not embark upon projects with the necessary enthusiasm and energy needed to succeed.

Using multiple methods of estimating ensures that when the results of the different methods are compared, a consensus is achieved of the likely out-turn costs.

The recognized methods of estimating are:

- parametric;

- analytical (detailed, grass-roots, bottom-up);

- analogous.

Within these methodologies there are naturally alternative models, for example PRICE Systems has two parametric models for software estimating: COCOMO II and TruePlanning Software model. These are both parametric mathematical models, but the cost research was performed by two independent organizations. They can therefore be considered independent methods of estimating.

Consider Figure 9.1: in Scenario One the result of the estimates are scattered, consequently we have very little confidence in the estimates. It is not possible to establish which method has determined the most accurate estimate. In Scenario Two, fortunately the estimates are grouped or clustered. In this case it is fair to suggest the Cost Engineer will have confidence to present the result of the analysis.

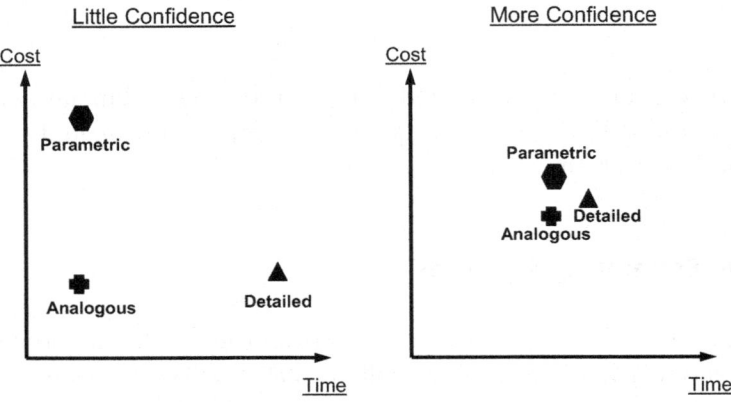

Figure 9.1   Multiple estimating methods

Top-down and Bottom-up

The analytical, bottom-up approach or detailed estimate is just that – detailed. However, detail does not necessarily bring confidence in the estimate. Constructing a bottom-up estimate is a difficult process and requires a great deal of organization if duplications and omissions are to be avoided. Unless there is plenty of discipline in the estimate it is possible for two separate departments to assume that they are both doing an activity or for both to assume that the other is doing it. However, this process does deliver buy-in and commitment to the estimate from those who contributed. For this reason management like the technique as they believe they will be able to hold project staff to the resulting resources if the bid is successful.

The advantage of top-down estimating is that it takes a holistic approach to the problem. The cost models consider the total problem, so items such as hardware/software integration resources are not overlooked or inadvertently added twice by different departments.

Constructing cost models requires a specific statement about the boundary of the cost model. For example, a hardware cost model specifically excludes software; there is no doubt of the boundary which can be confusing in some bottom-up cost estimates. This again is an advantage of the top-down parametric approach.

The disadvantage of the early parametric cost models is the challenge to achieve buy-in from departments that were not directly responsible for the resource estimate. The third generation parametric models are activity- and resources-based to enable departments to see their allocated resources and agree or modify them.

Considering the three recognized methodologies of cost estimating – parametric, analogous and analytical – the most inefficient method must be the latter. Analytical estimating, also known as detailed or bottom-up estimating, is a time consuming process. It consumes the time of the senior members of staff, experienced and trusted to do the estimate, just when the most skilled staff are needed to consider the management strategy for winning the bid and writing the technical solution.

Proactive Detailed Estimates

The detailed estimating process generally starts with the generation of a breakdown structure. This can change from industry to industry, but can take the form of a Cost Breakdown Structure (CBS) or a Work Breakdown Structure (WBS) or a combination of the two. The CBS is aligned with the accounting or financial control systems of the organization and will consider the categories of labour (engineering, manufacturing, and so on) and non-labour (material, printing, travel, and so on). The WBS is likely to be influenced by the project which is being considered. The WBS will terminate in a Work Package (WP), which is generally created at the same time as the estimates. The Work Package will stipulate:

- unique Work Package number;

- name of project;

- project phase;
- title;
- WP Manager;
- contractor name;
- issue number of WP;
- issue date of WP;
- planned start date;
- planned end date;
- precursor activities;
- deliverables;
- the object of the piece of work;
- the aim;
- items excluded;
- interfaces to this WP;
- task descriptions and details of the activities to be conducted.

When this process has been completed the author will have a good knowledge of the magnitude of the problem to be solved and the likely resources needed to solve it.

The format of the detailed estimate will change from industry to industry, but the important features are a time-phased estimate of individual resources, considering not just the immediate department of the WP owner, but also the non-labour that will be required to complete the work successfully.

This process has an undeniable advantage, which is the obligation to conduct the work within the resources estimated, if the WP owner has thought about the work necessary and then committed to completing that work with the resources stipulated. This buy-in to the estimate is reassuring to the senior management of the organization as it means that their chosen staff have been used to apply their best judgement in the creation of the estimate or bid with which they are being presented.

So what can be done to make this estimating methodology more efficient? Proactive Estimating!

It is highly likely that businesses will build on their experiences. A good successful business will develop a Business Plan which will continue to build on its historical strengths. As such it is highly likely that within a detailed estimating environment, staff are asked to estimate the same types of activities for the same types of technology repeatedly. This has to be inefficient; why not capture the thought processes that they repeat and store those thought processes in an estimating tool. This is the inspiration behind proactive estimating.

The desirable outcome of proactive estimating is to stop asking staff to estimate their activities and to present them with an estimate based upon a mathematical formulation of their thought process. For example, if the head of the drawing or drafting department is asked, 'how do you estimate the resources needed for a project?', they might provide a rule of thumb or little black book which contains their unique and consistent approach to estimating. Perhaps they assess the number of drawings required and allow 40 hours per drawing on average, plus 20 hours of computer time for the CAD package costs. If this rule of thumb can be captured and demonstrated to be the basis of all future drawing office estimates, this would save time and effort. In future the request from the Cost Engineering department would not be 'please provide an estimate', but 'how many drawings will you require?'

Having captured the Cost Driver in this way, an estimate can be presented for signature based on the original Cost Estimating Relationship. This is an altogether more streamlined process and capturing the knowledge ensures constancy.

Applying a third generation cost estimating model to this can change an estimating process from reactive – checking that estimates are correct logically and arithmetically – to proactive, where the Cost Engineering department generates activities and populates them with resources based on previously agreed parametric models that have been calibrated to meet the needs of the functions, like those of the drafting or drawing department.

Estimate Detail

The level of detail to which a system is broken down has a bearing on the level of accuracy of the resulting estimate. This theory is known as 'compounding errors' (Figure 9.2).

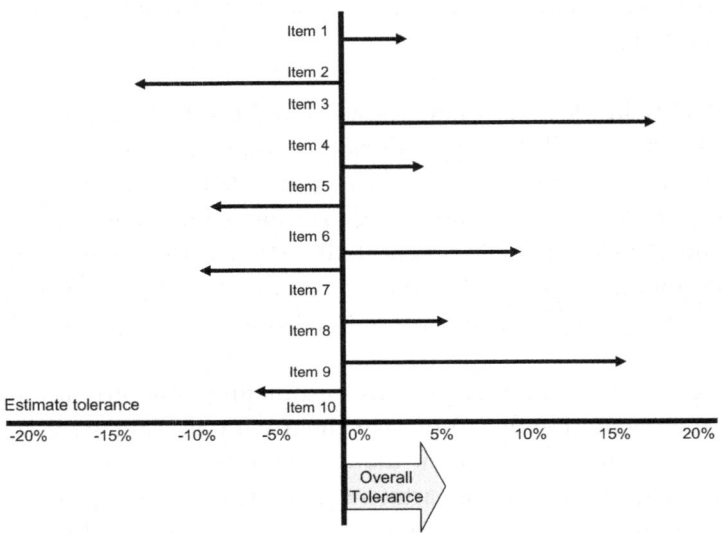

Figure 9.2   Compounding errors in estimating

This means that when the systems are broken down into smaller pieces, although the errors of those individual items will be the same, it is more likely that the accuracy of the overall system tolerance is narrowed due to the bias towards a higher population being correct. Figure 9.3 portrays this situation with earlier estimates being simple architectures with wide tolerance while, as time moves on, the architecture becomes known in more detail and the accuracy is increased.

Consistent and Repeatable

A software system such as that utilized by parametric Cost Engineers is an enabler. It is just a tool, but as such it provides a structure for the operation of Cost Engineering tasks. When the types of tasks being undertaken are repetitive, this lends itself to computerization. This in turn enables quality assurance of the deliverable.

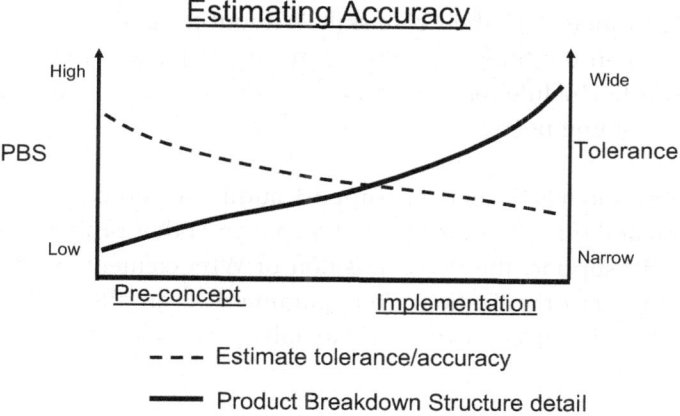

Figure 9.3 PBS detail and accuracy

Simply, parametric Cost Engineers are associated with estimating tasks. Once the cost research has produced a model, this can be broken down into: (a) gathering inputs to model, and (b) calculating and reviewing outputs of the model prior to reporting.

Cost Engineers have much to offer in the wider context of a bid or proposal. The nature of the work they conduct while estimating means that they have a unique perspective of the whole organization rather than just their own function. Adopting a parametric model frees the Cost Engineers to contribute to the bid or proposal and bring a unique and different perspective on the process. When time and resources permit, Cost Engineers can comment authoritatively on many aspects of the business from project management to production. For example, Cost Engineers can challenge, make or buy strategies, designs, support strategies, technologies and so forth.

A smaller number of staff will be focused on the building of the cost models. With the systems available today, this becomes a more structured and manageable process. The usability of the models created is repetitive and separates the cost model creation from the cost model usage.

Prime contractors in industry talk about *core technology areas*. They are content to let the fabrication and machining technologies go to a subject-matter expert, but they hold onto a core skill such as design, systems integration and high technology. In the same way, cost researchers can consider the building or creation of cost models as their core skill. Without these skills Cost Engineers are no better than the Integrated Project Teams (IPTs) or project staff at cost modelling. Cost, schedule and performance modelling is the unique competence that falls to Cost Engineers.

Parametric models, moreover, support audit trail processes through the identification and tracking of the project owner and other project stakeholders. Thus they will support the decomposition of WBS elements or the division of work within an organization. Thus parametrics can allow the allocation, management and collection of work to fall to specific specialists or team members.

Quick Response

As discussed above, when a computerized system is applied, it naturally results in a more streamlined, quality assured process. This condensing and focusing of the staff on the task will result in fewer resources being required.

Parametrics enable users to archive past projects or estimates in a unified knowledge base. This enables users to classify and define archive estimates as well as search and query archived data for reuse and forecasting.

Furthermore, shorter cycle time for tasks will be achieved when the input data requests are computerized.

Historical Trend Analysis (HTA)

The United Kingdom (UK) Ministry of Defence (MoD) is responsible for the acquisition of capabilities for the armed forces. In simple terms, they are responsible for the purchase of all military equipment from hand guns to aircraft carriers. When they began the SMART procurement initiative several years ago, amongst other things, the desire was to ensure that more of these purchases were completed on time and to budgeted cost. In the last few years, when the National Audit Office (NAO) has produced the Major Projects Report (MPR), it has been publicly acknowledged that some of these purchases are experiencing problems.

It has become widely accepted in many circles that these cost and schedule problems occur due to the unrealistic optimism in the initial estimates. Cost and schedules for new acquisitions are under pressure to conform to the budget that is available to ensure acceptance.

This is not to criticize the independence of the Cost Engineers in the UK MOD, which is the organization charged with estimating new equipment. Nor the desperate efforts of the Team Leaders in the Integrated Project Teams (IPTs) charged with the task of acquiring the equipment. But the environment is such that there is optimism and confident belief within the industry that the equipment will be cheaper than anticipated.

However the Investment Approvals Board (IAB), whose job it is to oversee the approval of projects, is trying a new 'weapon' to tackle the problem and provide quality assurance – Historical Trend Analysis (HTA). This is a very simple concept that has been utilized in industry for at least 15 years. One of the first practitioners of HTA was Mr Darryl Webb,[1] who used parametric cost models to normalize historical project costs and investigate trends. In his work he demonstrated a clear pattern of increasing project technology when considered over time (see Figure 9.4). This technique has been used successfully to extrapolate into the future and predict future technologies and the associated costs and schedules of future systems.

When run backwards, parametric models normalize historical project costs into a normalized cost density: an implicit numerical value representing the technology of a product and productivity of the organization that manufactured it. It is independent of economics, production rate, production quantity and currency and can be used to track technology trends relative to performance or productivity of companies relative to one another.

Figure 9.4 shows fighter aircraft systems plotted over time. The Coefficient of Determination (R^2) shows that there is a reasonable straight line correlation between these normalized cost densities and their associated In-Service Date (ISD). This technique is used to predict a future cost density, which together with other project specific parameters, is used to determine if the budget is realistic.

---

1   PRICE Systems Symposium, 1990 Varese, Italy, D. Webb, *Cost Complexity Forecasting Historical Trends of Major Systems*.

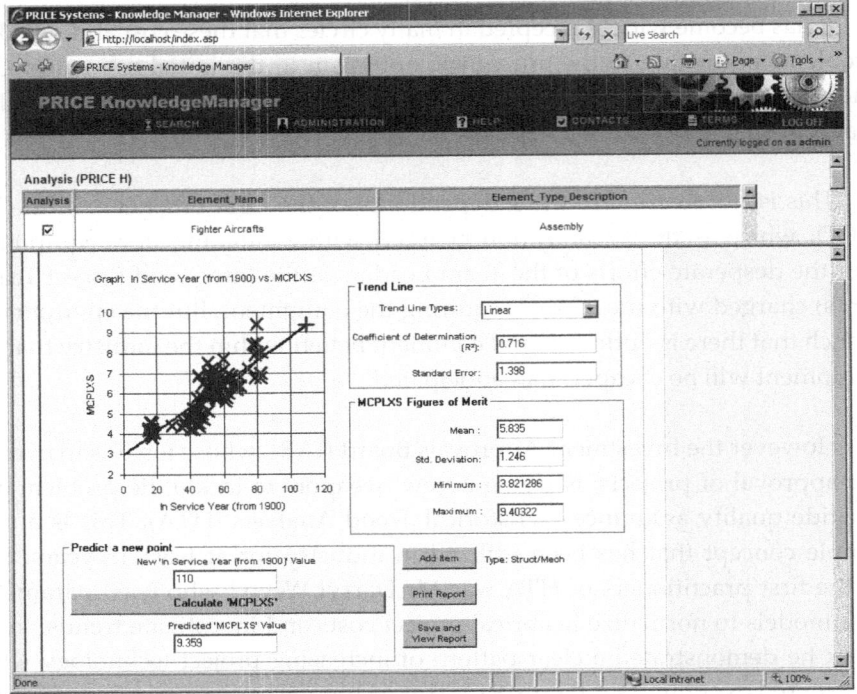

Figure 9.4 Fighter aircraft Historical Trend Analysis

The MoD has issued a SMART Approvals Guidance which recognizes this work as an independent and self-determining means of creating a comparator of their other estimating methods. The largest projects are now required to produce a comparator for their initial Gate submissions. Ultimately, such a comparator will be required for Main Gate submissions and medium-sized projects. This means that during the preparation of business cases, projects will look for comparable systems for the acquisition they are going to make and extrapolate into the future to give the budget and project estimate context. This can be achieved simply using knowledge bases that have been populated by the MoD based on actual costs of previously completed projects.

This just leaves the role of industry in the process. Clearly, if proposals to the MoD are to be believed, then an HTA accompanying the proposal will put the proposal in a favourable light.

Model Validation

The US Parametric Estimating Initiative (PEI) sponsored by the Defence Contract Audit Agency (DCAA)[2] and the aerospace industry has stimulated interest in using parametric cost estimating techniques as the sole Basis of Estimate (BOE) for proposal pricing.

Industry and Government practitioners have used parametric techniques to perform a variety of applications, such as independent cost estimates and trade studies. However, industry's use of parametric techniques as a BOE on proposals submitted to the Government or higher tier contractors has been relucant. This was a result of:

- a lack of awareness or understanding of parametrics, both within industry and the Government acquisition community;

- perceptions that regulatory barriers existed; and

- limited examples of actual proposal applications.

Industry and Government parametric practitioners asserted that proposal preparation, evaluation, and negotiation costs and cycle time could be reduced considerably through the increased use of parametric estimating. These practitioners also stated that these benefits could be achieved while maintaining the quality of the estimates produced.

In December 1995, the Commander of the Defence Contract Management Command (DCMC) and the Director of the Defence Contract Audit Agency (DCAA) sponsored a Reinvention Laboratory to evaluate the use of parametric estimating on proposals. Thirteen companies established Integrated Product Teams (IPTs) to test the use of these techniques on proposals.

The Parametric Estimating Reinvention Laboratory objectives included identifying:

- parametric applications to use as a BOE;

- specific barriers precluding the expanded use of parametric techniques and procedures used to address these issues;

---
2 www.dcaa.mil.

- requirements of a valid parametric estimating system;

- resources needed to support the implementation of parametrics; and

- benefits that could be achieved using parametric techniques as a BOE.

In addition, the Laboratory objectives included developing case studies based on the IPTs' best practices and lessons learned for use in developing the *Parametric Estimating Handbook*, and in establishing a formal certified training course.

To address the Laboratory objectives, the participants implemented a variety of parametric applications including Cost Estimating Relationships (CERs), company-developed models and commercial models. Of the commercial models, only PRICE Systems models were utilized by the industrial participants; no other commercial models were adopted.

The industry participants were able to negotiate contracts with their customers for proposals based on parametric techniques. In several of the test cases proposal preparation, evaluation and negotiation costs and cycle time were reduced between 50 and 80 per cent when parametric techniques were used as the primary basis of estimate. In addition, the IPTs found that the accuracy of the estimates was maintained or improved because of the increased use of historical data, thus demonstrating a quality assured estimating system.

The Laboratory results also demonstrated that when properly implemented and used, parametric estimating complies with the US Government procurement laws and regulations including the Truth in Negotiations Act (TINA), the Federal Acquisition Regulation (FAR), and Cost Accounting Standards (CAS). Case studies and examples were developed based on the IPTs' best practices and lessons learned, and they are incorporated into the *Parametric Estimating Handbook*.[3]

At the time, Mrs Eleanor Spector, Director of Defence Procurement, issued a letter in August 1995 to DoD component acquisition directors. In this letter

---

3   The Second Edition of the *Parametric Estimating Handbook*, issued in the Spring of 1999, can be obtained from the International Society of Parametric Analysts (ISPA) website at <www.ispa-cost.org>. It is also available on the Department of Defense (DoD) Acquisition Deskbook.

Mrs Spector states: 'I fully support the use of properly calibrated and validated parametric cost estimating techniques on proposals submitted to the DoD, and I encourage your enthusiastic support.' In January 1997, Ms Darleen Druyun, Principal Deputy Assistant Secretary for Acquisition and Management for the Department of the Air Force, issued a letter to the Air Force acquisition centres encouraging the expanded use of parametrics. In addition, there has been tremendous support for parametrics at the executive levels of the Army, Navy, DCMC, DCAA and NASA. As a result of such top-level commitment in the US, companies and their Government customers are more confident that parametric techniques can serve as an acceptable BOE in firm business proposals.

Achieving Concurrent Estimating with Client/Server Capability

Working together can mean many things: in the context of Cost Engineering it refers to Cost Engineering staff working together as a cohesive, efficient body. But it also means the customer working with their supplier to help resolve differences and difficulties with regards to cost, schedule and productivity.

Parametric models can be stored as XML files, making them light-weight and portable, thus promoting transmission and sharing between users. Further, Government organizations can effectively promote supply chain collaboration through the development and management of models between customers, suppliers, sub-contractors and partners.

A parametric model is able to identify projects as a baseline and enforce configuration management best practices to track and manage changes to the baseline. Further, users can be guided through organizational workflow to amend, edit or alter baselined breakdown structures. Parametric models can also support a peer review tick box. The reviewer's comments, user identification and timestamp will be stored with the project. Because many organizations lack the discipline to hold peer reviews of cost estimates that are produced prior to them being vetted by senior management or submitted to the customer, these peer review tick boxes prompt an opportunity for estimating error to be identified and corrected before embarrassing explanations are required to be given to the ultimate recipient of the work.

Collaborative Networked System

A parametric framework enables different types of cost models to exist within a single, unified estimating system. The various elements within the project being considered can therefore be estimated using the most appropriate model. Cost models can be detailed, bottom-up estimating models or top-down, parametric or Bayesian models.

Such a framework can provide multiple stakeholders the ability to organize and view data that is most relevant to their function or activities. It is a single workspace for the lifecycle of an estimate. The framework enables the convergence of multiple project perspectives into a single integrated project view. Project managers, cost analysts, engineers, finance and pricing groups are able to view and manage project data independently while the model tracks and maintains the integrity of the total project structure and properties. Thus the cost framework improves collaboration between stakeholders while ensuring estimating quality, consistency and coherency across each phase of a program (see Figure 9.5).

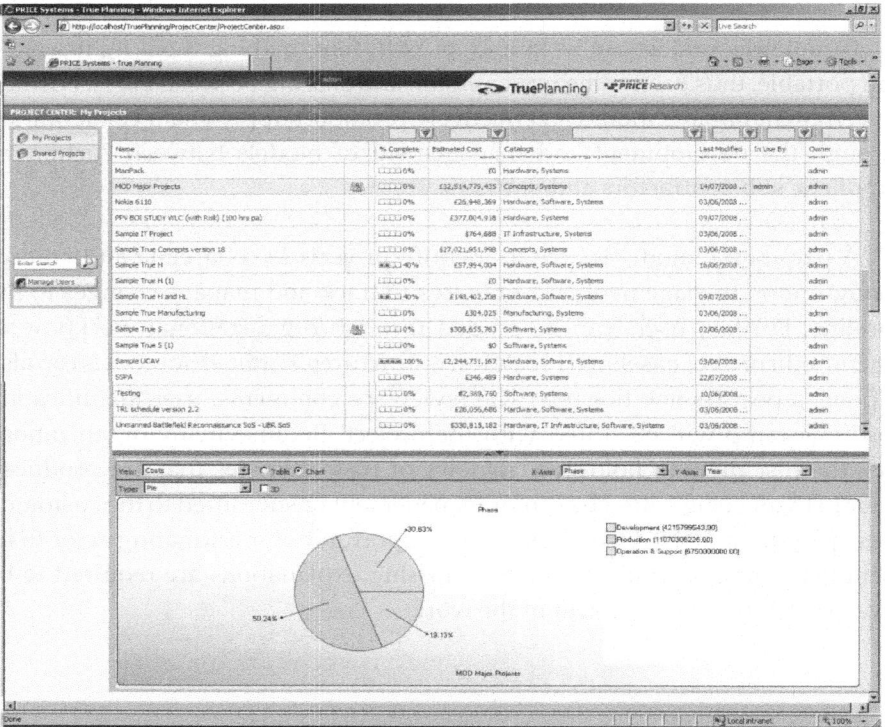

Figure 9.5   Client/Server Administration screen by PRICE Systems

In a client/server installation it is advisable to identify each user uniquely when they log in. This identification is additional to an organization's existing identification protocols to enforce data rights, access and security. User identification within a client/server environment enables a user's classes and groups to be created and managed. Organization and identification of users then enables the organization to assign and track work to be performed by the various users.

The Product Breakdown Structure can be decomposed across Cost Engineering based on the experience and expertise of the members of the team. The system allows for the allocation of work to specific team members. Figure 9.6 shows an example of an automobile software estimate being shared by specified staff. Cost Engineers might specialize in software, IT systems or hardware estimating.

Third generation parametric models are capable of using multiple estimating methodologies in the same estimate. For example, TruePlanning Software and TruePlanning COCOMO model are two methods for estimating software. For hardware estimating TruePlanning Hardware model, COSYSMO, TruePlanning Information Technology model and any models implemented as part of a conceptual model will provide multiple models for hardware estimating.

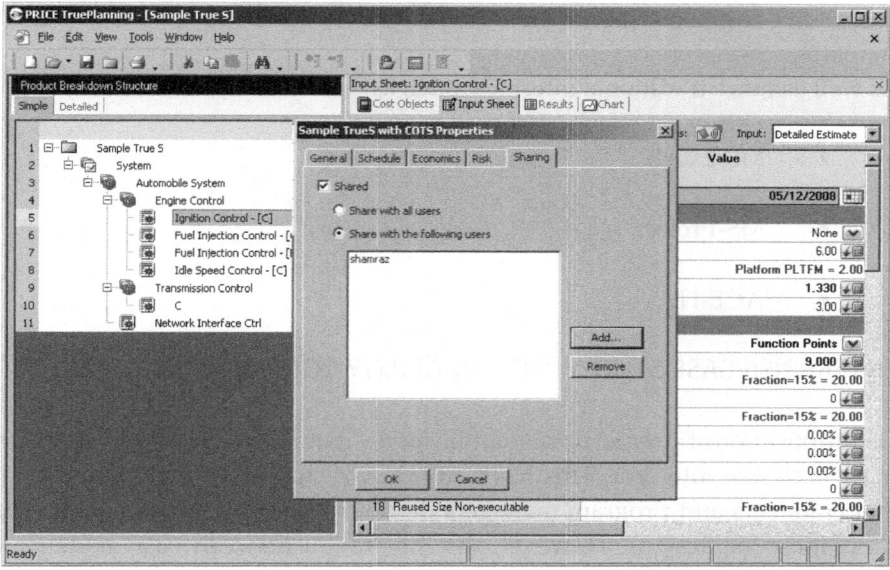

Figure 9.6 **Client/Server security**

Enterprise Connectivity

Modern third generation cost models use web services technology to enable enterprise applications to access the power of the parametric cost and schedule estimating engine. This provides estimating capability to design optimization tools to enable the analysis of stress and weight models to be combined with cost estimates, thus ensuring that the design is optimized in terms of not just engineering, but also cost.

There are many requirements for enterprise connectivity in Cost Engineering. Cost Engineers cannot conduct their science in isolation, they need assistance from many other sources. It is common for a Cost Engineer to interface with software engineers, accountants, systems engineers, schedule staff, project and bid managers, weight engineers, contracts staff, designers and more. As such they need to be able to communicate with these staff on the same intellectual level, but to make this effective and error free the estimating systems should be communicating directly with the systems used by these professionals.

Parametric tools offer a number of engineering interfaces for design optimization. For example:

- Phoenix ModelCenter 7.0;
- Engineous iSIGHT-FD.

As well as more traditional interfaces such as:

- MS-Excel;
- MS-Project;
- ACE-IT.

ENTERPRISE CASE STUDY – AFFORDABILITY COMPANION

One application of an enterprise solution for designers is the direct interpretation of their design into cost. This has a number of challenges including the fact that engineers and program managers do not routinely perform affordability simulations of designs. However, the integration between cost models and engineering design tools is key to assessing and managing costs early in the design process, before changes are costly. In this context the absolute cost is not the most important consideration. What we need to determine is the relative cost of one design compared to another. If a baseline cost estimate is quickly determined, any change of design can be measured as an increase or decrease relative to that baseline.

In organizations where cost estimating and engineering design tools are not integrated, it is difficult and time consuming to update the cost estimates as new information is received and provide feedback to the design engineer. Conversely, if integration is provided, cost has equal status with performance and design criteria, ensuring that when early design features are assessed, their cost is also included in the decisions.

Ideally, cost estimates need to be conducted quickly, the designer does not want to know the cost of including a feature in four weeks' time. Interoperability between engineering design and cost estimation systems is required to provide real time, predictive cost assessments in response to changes in the design process. The result leads to an optimized or 'best value' solution.

Hence an enterprise-capable estimating system is an answer, woven into the organization's systems to ensure an estimate when required. For the design engineers this dream is a reality with an integrated environment; designers and Cost Engineers may collaborate virtually and in real time to understand the impact of design changes on cost estimates. In addition, project, program and business development managers have instant access to the results of the iterations to improve their decision-making.

Parametric estimating models have proven technology to make the integration of cost and engineering design models possible: for example, the integration between Pro Engineer and the True Planning Cost model. In Figure 9.7 it is possible to see the three-dimensional representation of a piston, crank and housing. By using the Affordability Companion it is possible for the designer to acquire an estimate instantly.

Following the selection of the Affordability Companion the designer is prompted for a few answers, such as the quantity to be manufactured and when manufacture will commence. Once these have been completed, the system translates the Bill of Materials (BOM) in the Computer Aided Design (CAD) system into a parametric model in the form of a Product Breakdown Structure (PBS). In the error free information transfer the interface will also

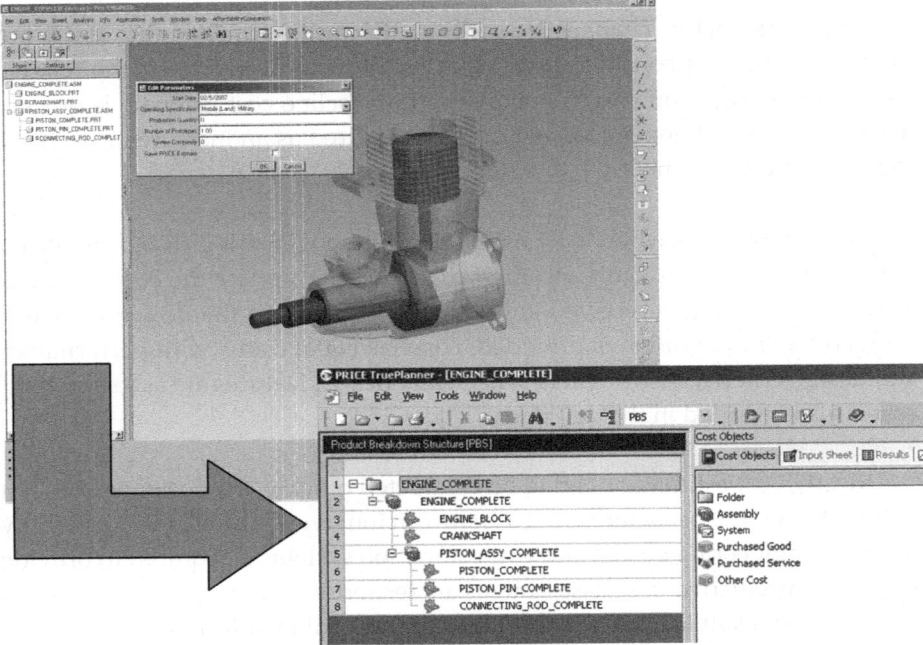

Figure 9.7   Computer Aided Design (CAD) and parametric estimating integration

move the material type, dimensions, density, number of features, and so on. The parametric systems will determine the likely manufacturing process to be used and estimate the associated cost. This is then communicated back to the CAD tool in the form of tables and graphics to set a baseline for this design.

It is now possible for the designer to consider alternative designs, which might be lighter, quicker, stronger, more powerful, and so on, and to establish how the new designs will influence the costs of the product. This empowers engineers to think about cost in the design phase, and to be able to perform 'what if' and risk analysis, since a change later in the development of the design is expensive and can affect multiple systems.

The enterprise-capable parametric model will also ensure that a cost estimation is included for integration and test activities on various configurations in real time. Ultimately, this allows engineers to conduct real-time trade studies from within Pro Engineer for evaluation of Total Cost of Ownership (TCO) and Cost as An Independent Variable (CAIV) analysis.

Estimated cost can be a real-time, truly independent variable within a modelling and simulation environment.

Summary

In this chapter on quality assurance you have learnt that:

- Multiple methods of estimating within one framework will provide a consensus view on the cost and schedule estimate.

- Proactive estimating driven by a parametric model can provide speed and commitment to an estimated resource.

- Historical Trend Analysis (HTA) can be used to verify cost estimates based on past experience.

- The parametric estimating initiative has stimulated interest in using parametrics as a basis of estimate for proposal pricing.

- Automated error-free transfer of information between other systems and a parametric model ensures quality assurance; it encourages what-if analysis and alternative solutions.

10

How to ... Estimate Through Life

The definition of Through Life Estimating (TLE) is 'the ability to generate repeatable cost and schedule estimates in a way that is traceable and scalable throughout the life of a program utilizing the most appropriate methodology at each stage of the estimate.'

This is a challenging objective. The very nature of projects means that they evolve over time. In the case of aerospace and defence projects this can mean many years. In other industries the time period may be shorter, but the challenge is not diminished due to the number of alternative designs that can be considered over the lifetime.

Moreover, swapping the estimating tool is like swapping the operating systems of the computer. The fundamental basis of the estimate has been eroded and thus the confidence has been impaired. But more seriously, it is difficult or not possible to trace the differences between the estimate created with one estimating tool and that produced by another.

As a philosophy the concept of estimating through life throws up many obstacles – from tools, to staff through to the estimating methodology employed. Maintaining a consistent professional estimating staff with appropriate knowledge of the industry, product and estimating processes necessary is a constant challenge for any organization over a short period. When projects span decades, as in the case of aerospace or defence projects, the problem is compounded. It means that providing a consistent estimating service to customers is difficult. Knowing the internal or external customer needs, and providing the service, deliverables and processes that satisfy those needs is a challenge.

Using the appropriate estimating methods during the maturing of the product or service such that the methodology matches the level of detail and information that is available at each stage of the life-cycle is never easy. In the past, spreadsheets provided a flexible solution, but they are time consuming. Staff spend much of their time learning how to program macros or use advanced functions, rather than concentrating on the task in hand, which is estimating. This in turn leads to complications in the ability to re-estimate as requirements emerge, and being responsive to the changes in the end customer's needs as well as those of the internal management.

Through Life Estimating involves ensuring quality assurance of estimates by using independent estimating methods to verify the estimate. Usually one method is difficult enough to assure under bidding conditions. It would certainly ease the situation during a frantic proposal if technology was available to establish a collaborative environment to communicate effectively and efficiently without the need for excessive paper. In such circumstances it will be more efficient for the Cost Engineer to be proactive and propose an estimate to the organizational functions rather than cross-checking the estimates produced by the organization.

As the life-cycle of the product changes and different estimating methodologies become relevant, the ability to use validated and verified estimating models through the life of the project without swapping the estimating environment is attractive. Likewise, the ability to capture historical data consistently will lead to the monitoring of efficiency changes and the justification of any investments made throughout the project life. Finally, the integration of the estimating suite with the tools used throughout the life-cycle of a program will provide an efficient environment in which to work: integration stemming from CAD and design optimization tools through to operating and support tools.

The diagram in Figure 10.1 demonstrates how it is possible to frame the challenge.

Over the lifetime of a program the level of detail and information about the program deliverable changes. At the beginning of the program, during the concept and assessment phase, little detail is available, and the significant focus is on finding revenue or a funding stream for the program. As such parametric or analogous estimating techniques are employed; these could be referred to as gross estimating methods.

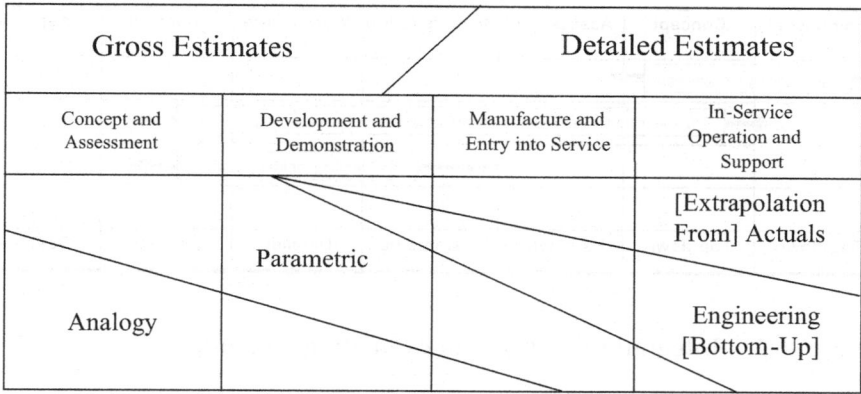

Figure 10.1 Through Life Estimating methodologies

As the program matures, the level of information increases until detailed estimates are desired. During the phases of manufacturing and entry into service, bottom-up estimates are required and actual costs become available. These actual costs are substituted for the estimates that have been used previously. This is not a clean process and during the maturity of the program, hybrid estimates will occur which use a mixture of estimating methodologies.

Spreadsheets are an obvious choice for solving the Through Life Estimating problem. Due to their infinite flexibility they can be used to assist a Cost Engineer in the process of preparing an estimate using any methodology. But this flexibility is also their shortcoming, as they are not designed for cost estimating. They are used for budgeting, calendars, expense reports, inventories, invoices, purchase orders, evaluations, games, ledgers, tests, menus, reports, the list is endless. What is required is an estimating framework, as flexible as a spreadsheet in terms of its ability to accommodate different estimating methodologies, but designed specifically with cost estimating in mind.

A well-developed parametric model, such as TruePlanning (see Figure 10.2) is an example of such a framework with cost estimating methodologies appropriate to each phase of a program.

A third generation parametric cost model is an estimating framework which addresses Through Life Estimating challenges by providing a consistent approach to risk analysis, escalation, labour rates, input parameters, reports,

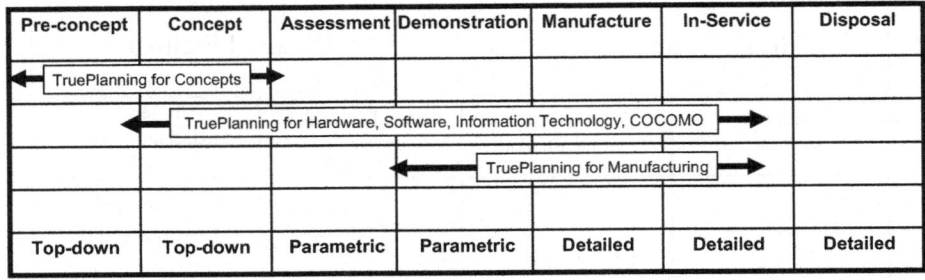

Figure 10.2   Integration of different estimating methods

and so on. It will involve a collaborative, client/server environment to enable the sharing of the paperless information throughout an organization:

- Using a parametric cost model will boost productivity since scalable cost modelling is available in portable form on a laptop or corporately adopted on an enterprise server installation. Through Life Estimating with a third generation parametric model will provide the ability to create custom cost models manually which are validated and verified by the organization. This will facilitate credible, traceable estimates for equipment, systems and Systems of Systems.

- Commercial parametric tools have sophisticated knowledge management features to enable the retention of corporate or individual knowledge. With the addition of an Affordability Companion, knowledge management is still further enhanced by interoperability with other enterprise tools such as design and design optimization tools.

- Tutor-based training can be augmented by web-based training to ensure that staff learn at their own pace and have the ability to revise their learning when required.

A Through Life Estimating solution provides a series of cost and schedule estimates as shown in Figure 10.3.

Through Life Estimating ensures that estimates can be reproduced as requirements emerge. This is dependent on the estimating framework

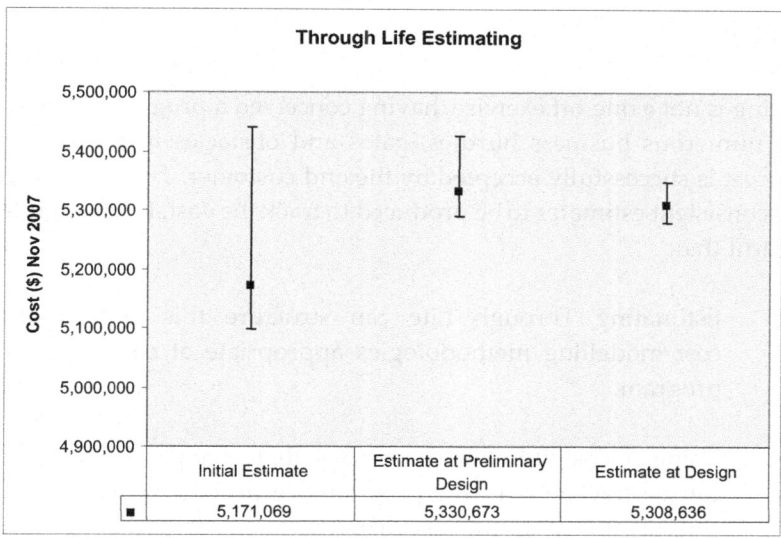

Figure 10.3   The desired outcome of Through Life Estimating

accommodating appropriate estimating methodologies throughout the life of a program, from cradle to grave.

Ideally, the Cost Engineers will have their skills maintained using web- and tutor-based training to professional standards to ensure a consistent professional estimating staff through the life of the project. This will enable them to provide a consistent estimating service to customers in terms of deliverables and process.

Cost models established in a collaborative environment ease communication effectiveness and encourage capturing historical data consistently, thereby helping monitor efficiency changes. Senior management and decision-makers can monitor the effect of their improvement initiatives, rather than pondering if the difference was caused by a change in estimating model.

Finally, the Through Life Estimating implemented in an open framework architecture provides easy validation and verification of estimating models. When implemented in an enterprise environment the model creates opportunities for integrating estimating with other Product Lifecycle Management (PLM) tools, thus saving time and resources.

Summary

Estimating is not a one-off exercise, having conceived a program of work there will be numerous business hurdles, gates and obstacles to pass before the system cost is successfully accepted by the end customer. These obstacles will require consistent estimates to be produced to track the cost. In this chapter you have learnt that:

- Estimating Through Life can structure this application with cost modelling methodologies appropriate at each phase of the program.

- Using a cost framework ensures that changes of requirements can be tracked over the program life providing justification in a consistent manner for cost changes.

- The problems of a single point failure, validation and verification, training and configuration management associated with spreadsheet models can be overcome by adopting third generation parametric models.

- Through Life Estimating encourages corporate knowledge retention formally leading to cost and time saving in the estimation of the next generation of your systems and programs.

11

How to ... Estimate Technology Maturity

During a recent customer support visit I was asked, 'Can parametric cost models be used to predict the cost of projects with an influence on Technology Readiness Level (TRL).' This lead to a discussion on the concept of TRL and this chapter documents some of the issues that were discussed and their solutions.

Technology Readiness Levels (TRL)

Technology Readiness Levels were conceived by NASA to provide an aid when quantifying the level of risk inherent in their programs on the basis of the novelty of the technologies being considered. Figure 11.1 encapsulates the idea behind the TRL scale.

At the top of the scale are mature technologies (TRL 8 or 9), which are proven in the field. They are well documented and are widely known and understood. At the bottom (TRL 1 or 2) are the immature technologies representing the challenge to the scientific community as they are not yet well enough understood to be practical for engineering application.

There are two reasons to investigate this issue. Technology makes the transition from the laboratory to the engineering workshop via two routes: technology push and technology pull. Technology push is scientist-driven and requires that pure research is conducted without the forethought of application. Once the technology is discovered, the scientific community take it upon themselves to find an application for their new discovery. Technology pull is the recognition, from the practical engineering perspective, that technology could overcome a shortfall in performance. A systematic search for technologies that can overcome that shortfall is then undertaken and a solution sought.

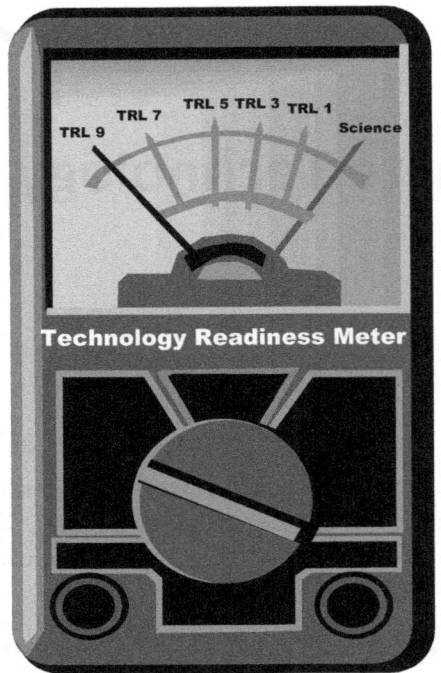

Technology Readiness Level (TRL):
9 = Proven technology
7 = Demonstrated technology
5 = Usable technology
3 = Applied Research
1 = Pure Research

Figure 11.1 The Technology Readiness Level (TRL) scale

The Costs

So what are the costs that are actually incurred when implementing technology at different TRL states? When considering the cost effect relating to TRL, all the common costs need to be subtracted from the calculations – for example, the cost of manufacturing the items which are not novel in nature; the other parts of the system. The simplest way to consider the cost effects of TRL is to approach the question from the perspective of a mature technology and consider what expenditure occurred during its transition from immature to mature.

There are several categories of cost to consider:

1. *Maturing the technology* – before the immature technologies can be put to reasonable use they need to be studied and investigated. There is non-recurring research and development funding needed including documentation of the technologies' physical properties and utility.

2. *Investing for the technology* – before the technology is applied there needs to be investment in capital facilities for that technology. Investment in new buildings, infrastructure and tooling needed to prepare for the new technology.

3. *Manufacturing the technology* – finally there is the actual manufacture of the technology itself. The labour and material required to produce this unique product.

This chapter will consider each of these categories in turn. However, it should be stressed that the Cost Drivers used to provide a solution are not working in isolation, but are combined together in the cost model to provide all the necessary influences.

We need to isolate the immature technology using the Product Breakdown Structure. The other mature technologies can be considered in the normal way once the immature technology has been isolated is this way.

Maturing the Technology

The weight and complexity input parameters in the cost model determine the magnitude of the data package that is required during the non-recurring phase of the project to enable the engineers to communicate their design to the manufacturers of the prototype. The prototypes are needed to ensure that the design will satisfy the requirements of the project. This data pack contains specifications, drawings, designs, Bill of Materials, and so forth – all the documentation necessary to produce the prototype.

The New Design and Design Repeat parameters will determine the uniqueness and the heritage of the design respectively; they determine the size of the data pack, the bundle of documents that are transferred from the design team to the manufacturing team to enable either hardware to be described in sufficient fashion that it can be accurately made or software described in enough detail that it can be coded into a programme. The Design Repeat parameter is not influenced by the technology maturity: it really reduces the data pack (the Bill of Materials, drawing count, material specifications, process and so forth) when considering multiple designs; for example, left-hand and right-hand designs. The New Design parameters do play a part, but only in establishing the relative proportion of the design that is considered novel, which is not a direct consequence of technology insertion but other design considerations such as two identical wings required to produce the same amount of lift resulting in repeat designs. If the item being modelled is mostly an established design from a previous product, then the New Design will be low. If the item is mostly a novel design with new technology then the New Design parameter will be high.

The significant parameter in all this is the Engineering Complexity input. The Engineering Complexity can be used to adjust the amount of resources required by considering the difficulty or scope of the design task. The effort required to produce the data pack is adjusted by the Engineering Complexity on the basis of two dimensions:

1. the skill of the engineers employed, and

2. the scope of the design.

The TRL is independent of the skill of the team; engineers who have just graduated will take longer to compile the data pack than seasoned engineers regardless of the scope of the task.

When considering two identical projects of the same weight, complexity, New Design and Design Repeat, the resulting resources are the same. However, the significant dimension is the scope of the design. As can be seen in Figure 11.2, the scope of the design effort varies from a simple modification to a state-of-the-art design.

These can be explained as follows:

- In a *simple modification* it is easy for the engineers to alter an existing data pack from a previous design; change the parameters, change the titles, alter the sizes, and so on.

- For an *extensive modification* more effort is required to realign the existing data pack, but as people say 'it is still easier to edit than to write a book'.

- When the non-recurring work is *New Design* then it is not possible or cost effective to modify a previous data pack and it is necessary to start from a blank canvas. All the documents must be generated within the data pack; however, because the technology is existing,

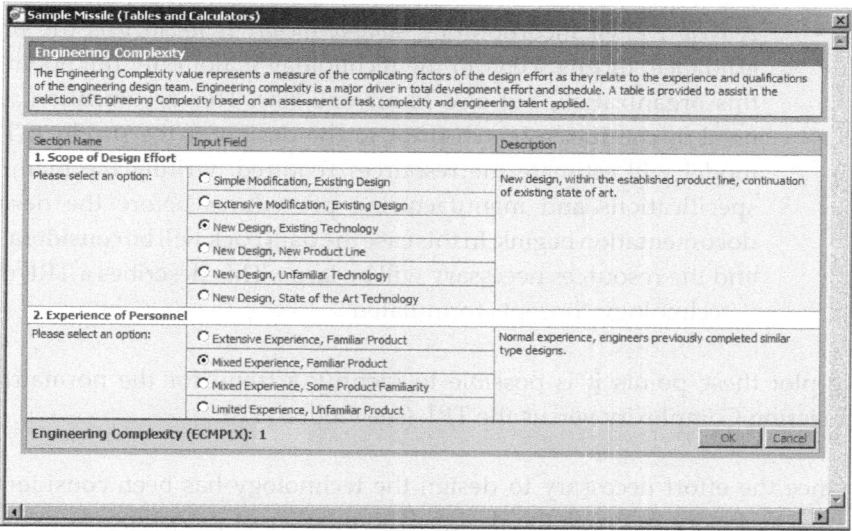

Figure 11.2   Engineering Complexity Calculator

references to company material specification and processes make the task focused on the application of these for the New Design. This explanation of New Design is also a good description of technologies in the TRL range of 6 to 9 – 'Proven Technology'.

- A *New Design* with a *New Product Line* will require more resources as a result of the novelty of the configuration. The technology is assumed to be known as above, so existing internal documentation can be referenced, but research is required into this particular configuration of the technology. This describes a TRL of 5 – 'components / breadboard validated in relevant environment'.

- A *New Design* with *unfamiliar technology* will require still more resources due to the need to research the technology employed. The technology is unfamiliar to the organization, therefore effort is required to research the data pack prior to considering the configuration of the technology in a design. In the context of TRL, the technology exists but is not mature in this particular organization. The inference therefore is that external assistance is necessary or that internal investment will be made to become familiar with this technology. This describes a TRL of 4 – 'components / breadboard validated in laboratory'.

- A *New Design* incorporating *state-of-the-art technology* is the final guidance. In this scenario the technology is not only unknown to this organization, but also to anyone else. This organization will need to conduct research prior to the design of the product. The model will estimate the resources required, writing the material specifications and manufacturing procedures before the design documentation begins. In this case the data pack will be considerable and the resources necessary will be large. This describes a TRL of 2 – 'technology concepts formulated'.

If we plot these points it is possible to identify a trend for the normalized Engineering Complexity versus the TRL (see Figure 11.3).

Once the effort necessary to design the technology has been considered, it is still necessary to prove the design. The number of prototypes is used to demonstrate that the design satisfies the requirements in full. This is achieved by the use of engineering models, test units and other common items that are qualified through acceptance tests during the development phase.

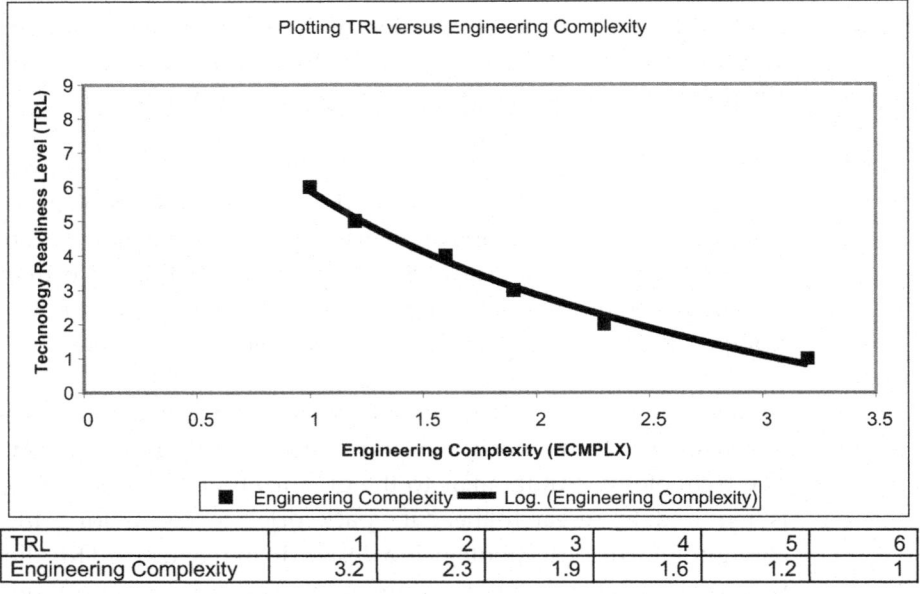

Figure 11.3 TRL versus Engineering Complexity

It is normal practice to validate and verify any maturing technology with prototypes. However, these are used in addition to those required to demonstrate the design. The prototypes needed to demonstrate the technology are a result of technology yield. The Prototype Support Adjustment Factor is a parameter used to adjust prototype manufacturing and development engineering costs when design / fabrication yield problems are anticipated. But it will also influence the costs when technology is not mature.

The nature of cost changes with variation in Prototype Support Adjustment Factor is twofold:

1. *Prototype manufacturing* – Prototype Support Adjustment Factor is a direct multiplier of this cost element; meaning that a value of 2 for Prototype Support Adjustment Factor will produce twice the prototype manufacturing cost of a Prototype Support Adjustment Factor value of 1.

2. *Development engineering* – Prototype Support Adjustment Factor causes a non-linear (generally a logarithmic) increase to the development engineering cost elements. This reflects the expectation that design yield problems are probably isolated to a subset of components of the assembly in question. In other words, multiple design approaches do not equate to multiple designs for every part of an assembly. A non-linear, logarithmic equation that is prototype-quantity sensitive conveys this reality.

For mature technologies (TRL 9) there is no need to alter the Prototype Support Adjustment Factor, but in a non-mature technology (TRL 1) the Prototype Support Adjustment Factor would need to be increased to reflect the potential for failed design paths resulting in the use of immature technology.

Investing for the Technology

Parametric models include the cost of the procurements (design and manufacture), of the tooling and the cost of support and maintenance of the tooling and test equipment. This cost is driven by the monthly production rate and the complexity of the technology being procured.

When a mature technology is being considered (TRL = 9), the development tooling and test equipment cost will be low due to the Engineering Complexity of the items. When immature technology (TRL = 1) is being considered within the solution, the development tooling and test equipment procurement will be high due to the Engineering Complexity. In both cases the production tooling and test equipment is assumed to be similar as by the time we get into production the technology will have matured to a state that enables production manufacture.

In the extreme it could be necessary to add development and production sets of tooling to indicate the design yield or failures which would litter the workshop when trying to mature the technology and develop the manufacturing methods.

The integration and test activity has Cost Drivers for the electronic and structural plans. This accounts for the test plans and procedures required to conduct the integration and testing of the items successfully. To ensure that the correct level of effort is calculated against the integration and test items, this input needs to be considered.

The cost of any civil infrastructure project will be excluded from this estimate. It might be necessary to add the cost of a new manufacturing facility to the estimate as a throughput or other cost.

Manufacturing the Technology

Parametric hardware models have parameters which are used to describe the technology manufactured within a project. Manufacturing Complexity is an obvious example, which is a technology index that has been constant since the early 1970s. However, experience has shown that the Manufacturing Complexity of systems has changed over time as the systems have adapted and changed to evade any threats in defence and competitive pressure in the aerospace market. The Manufacturing Complexity has remained constant as an empirical figure due to other parameters which influence costs of the technology. A simple example of this is the escalation tables which reflect the effects of inflation on the labour and material cost calculated. A more sophisticated parameter is the year of technology.

The PRICE System's models have a technology model built into them. This ensures that the effects of market forces can be reflected in the costs produced. In Figure 11.4 it is possible to see that at the time of the project (Time Now) a lower complexity solution will yield a cost benefit due to the abundance of skilled labour and material suppliers of this technology. But a higher complexity solution would result in a cost penalty due to the lack of supply when considering the labour and material suppliers available.

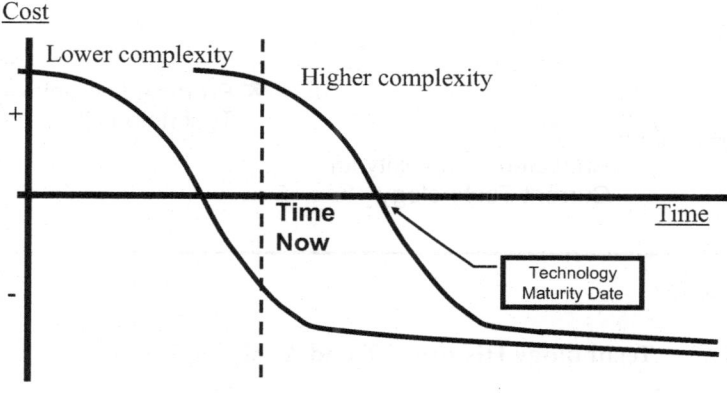

Figure 11.4 Technology Maturity model

The maturity of the technology has this effect. For mature technologies (TRL = 9) the Technology Maturity Date will have already occurred; the technology is mature at the point when the program requires its application. However, for immature technologies (TRL = 1) the Technology Maturity Date will still be in the future, indicating that a higher cost is appropriate.

Using Historical Trend Analysis (HTA) (Figure 11.5) it is possible to predict future technology costs. The equipment will have a complexity for current technologies and by plotting these by time, it is possible to predict the cost of the future, at present immature, technologies.

Modelling the Studies

To consider the process in more detail, the single estimate needs to be broken into several elements. When considering technology maturity in more detail, guidance becomes more problematic without knowledge of the actual technology program. This section will therefore provide guidance regarding the input parameters to be considered and how individual estimates should be tailored using precise understanding of the technology program being contemplated.

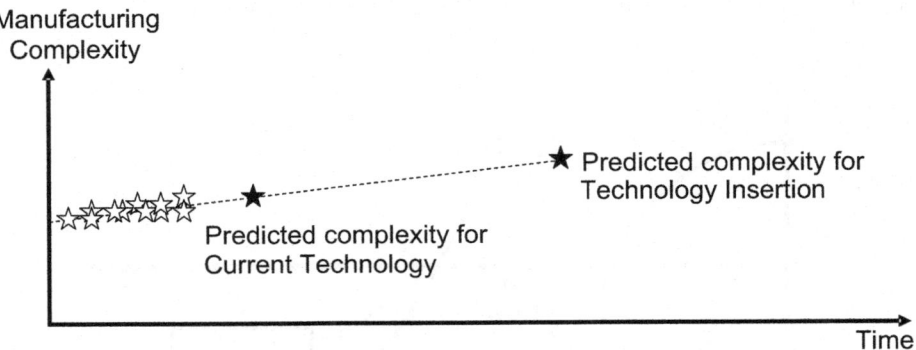

Figure 11.5 Technology Historical Trend Analysis (HTA)

Figure 11.6 describes the phases of work which lead to the maturity of technology that can ultimately be delivered in a product.

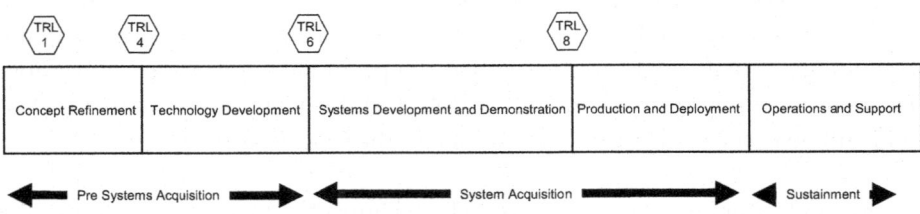

Figure 11.6 TRL phases

Parametric cost models can be used to estimate the different phases of work recognizing that these packages are discrete studies which can, at any time, be cancelled. Figure 11.7 is a typical program, with some studies making a contribution to the next phase of work, and some terminating after completion. The phases of work are focused upon, and aim to lead to, the maturity of software or hardware technologies. However, there is no guarantee that the technology will make it into the final product; the study might be successfully completed but prove too advanced to be of immediate use and be consigned to the shelf until other technologies catch up.

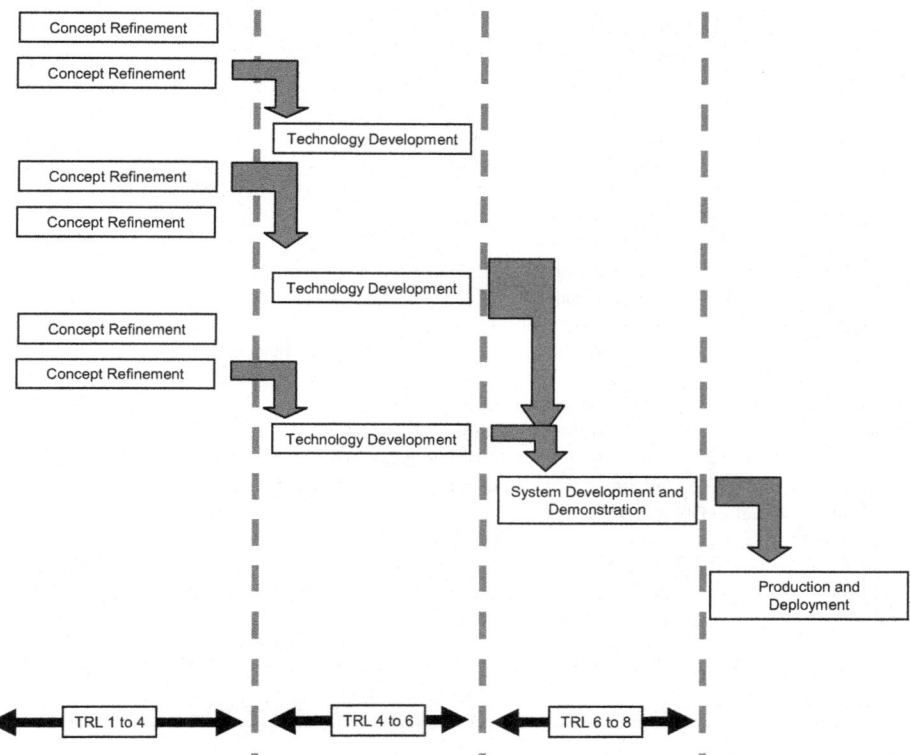

Figure 11.7 Progressive refinement and merger of technology and research

The Product Breakdown Structure (PBS) of the phases of work are shown in Figure 11.8. Within this estimate activities and appropriate resources are considered within an activity-based estimating framework. Each study has a system and assembly cost object to ensure that the study is self-contained with both management and technical oversight.

In the following sections, some considerations are given to the parameters in a detailed estimate.

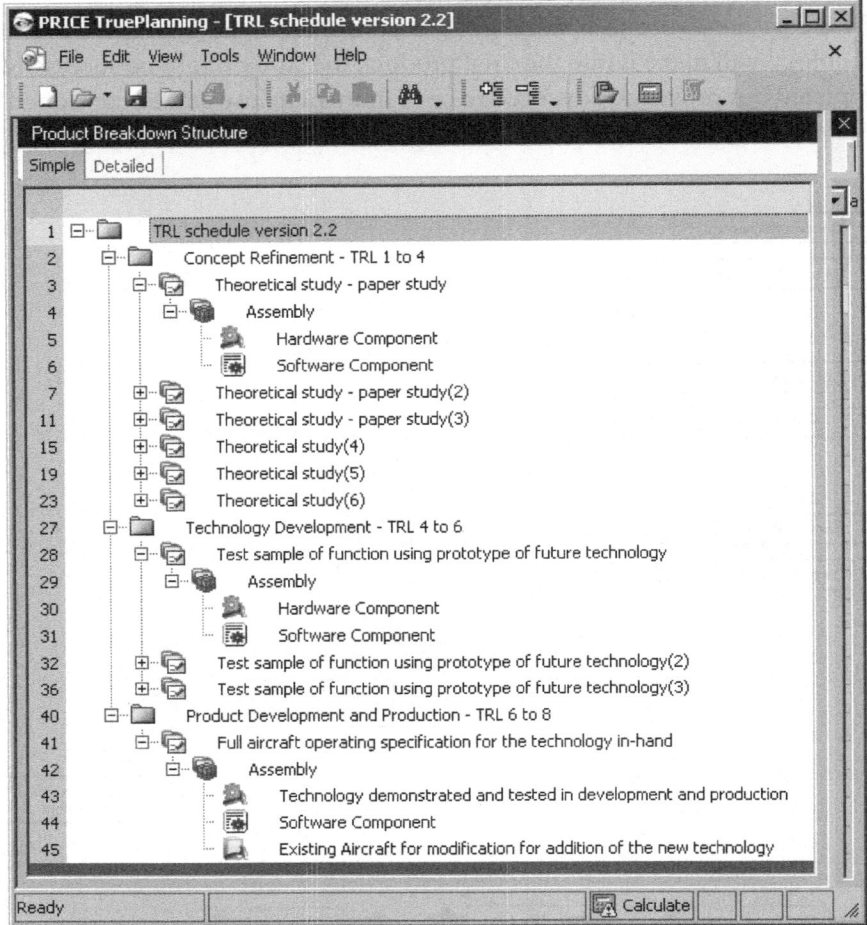

Figure 11.8 TRL Product Breakdown Structure (PBS)

QUANTITY OF PROTOTYPES

Parametrics can be used to estimate the cost of paper studies (non-hardware) for early TRL levels through to ultimate product development and production. To estimate the paper study, the input parameters are the same, but both the Production Quantity and Prototype Quantity are zero.

As the technology matures the research organization will begin demonstrating the technology in the laboratory through to the real environment. Prototypes can be calculated depending on the number and type of hardware to be produced. Fractions of a whole prototype can be entered into the model based on the hardware configuration and detail required to demonstrate the technology.[1] In the absence of a prototype calculator, an MS-Excel link can be used – an example is shown in Figure 11.9. Consider the number of brass-board models, engineering models, structural models, thermal models, and so on that might be required in each phase; it could be more than one.

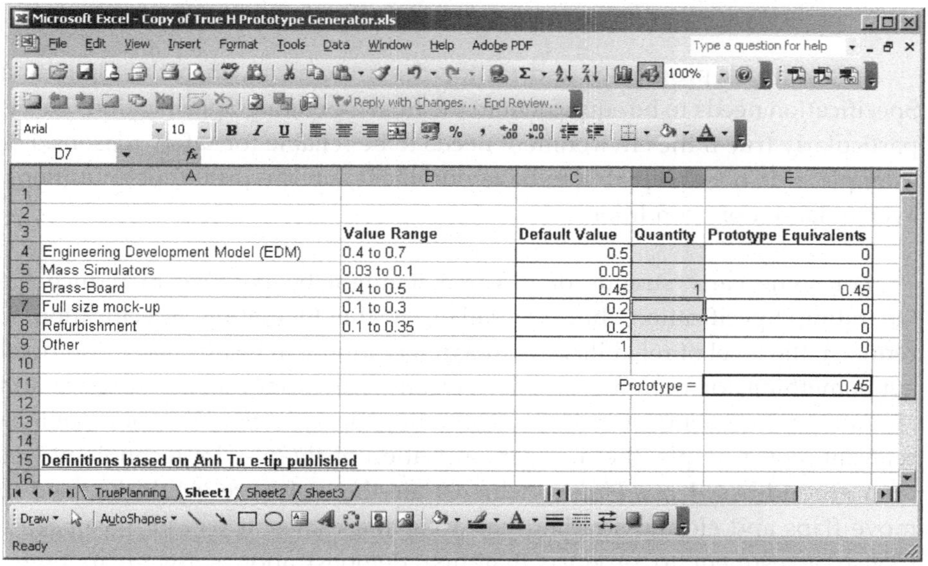

Figure 11.9 MS-Excel link calculating the number and type of prototypes

It is likely that the number of equivalent prototypes required will increase as the technology matures (see Table 11.1). Ultimately the technology will need to be proven in the full environment in which it will be deployed to gain certification and qualification status.

OPERATING SPECIFICATION

The early work will be primarily paper studies: pure research without application consisting of literature surveys, library searches and the advancement of fundamental principles that might have application within an industry. As such it will deliver no hardware or prototype.

---

1 Anh Tu, 'Prototype Equivalents', unpublished PRICE Systems research paper.

Table 11.1   Number of prototypes

Phase	TRL	Number of equivalent prototypes
Concept refinement	1 to 4	Nil to low
Technology development	4 to 6	Medium
System development and demonstration	6 to 8	High

As the work progresses and prototypes are created, then the Operating Specification needs to be equivalent to the final Operating Specification. This is particularly true if the environment needs to be reliable for safety reasons. For example, it is not acceptable to have unreliable, unsafe prototype munitions, even in laboratory conditions.

For some early strands of research it might be possible to reduce the Operating Specification, but remember, if the Operating Specification is reduced, the level of reliability and documentation will be affected – you don't get something for nothing. This is typically only applicable in a space and aircraft environment; at a lower Operating Specification it would not be relevant. For example, the first demonstration of fly-by-wire technology was on a ground-based *Iron-bird* to demonstrate that software would be used to move flaps and elevators rather than rods and cables. It would have been a foolish investment to produce that first demonstration software to flight-worthy quality.

NEW DESIGN

At the start of the Concept Refinement phase there will be a considerable level of novel design effort required. However, the aim of the technology maturity program is to progressively de-risk the technology and learn from the previous phase. As such, the level of new design should reduce as the technology matures (see Table 11.2).

ENGINEERING COMPLEXITY

The level of applicable Engineering Complexity has already been discussed. This would still apply in the detailed estimate, as shown in Table 11.3.

Table 11.2   New design

Phase	TRL	New design
Concept refinement	1 to 4	High
Technology development	4 to 6	Medium
System development and demonstration	6 to 8	Low

Table 11.3   Engineering Complexity

Phase	TRL	Engineering Complexity
Concept refinement	1 to 4	High
Technology development	4 to 6	Medium
System development and demonstration	6 to 8	Low

SOFTWARE

During the early phase of the technology maturity process, software can be a consideration. Initially, at TRL 1, software studies will not consider the production of code, the top-level design of the software will be sufficient at this phase. Consideration of the function of the software and review of the existing code libraries for these functions might be appropriate. In estimating, therefore, it is necessary to create a Worksheet Set that will remove all activities including and beyond the code and unit test activity within the software life-cycle.

A sample result is the output in Figure 11.10. The various studies are independent and stand alone. The results show the cost of each study spread over the appropriate time in the total program. In reality these studies would be conducted over much longer periods and could contribute to the learning of another study.

Summary

In engineering we constantly strive to develop new technologies to enhance our system and equipment to provide the best possible capability. But this effort is all in vein unless you have the ability to estimate the resources that it will take to bring the latest technology to the market. The best engineering has been left

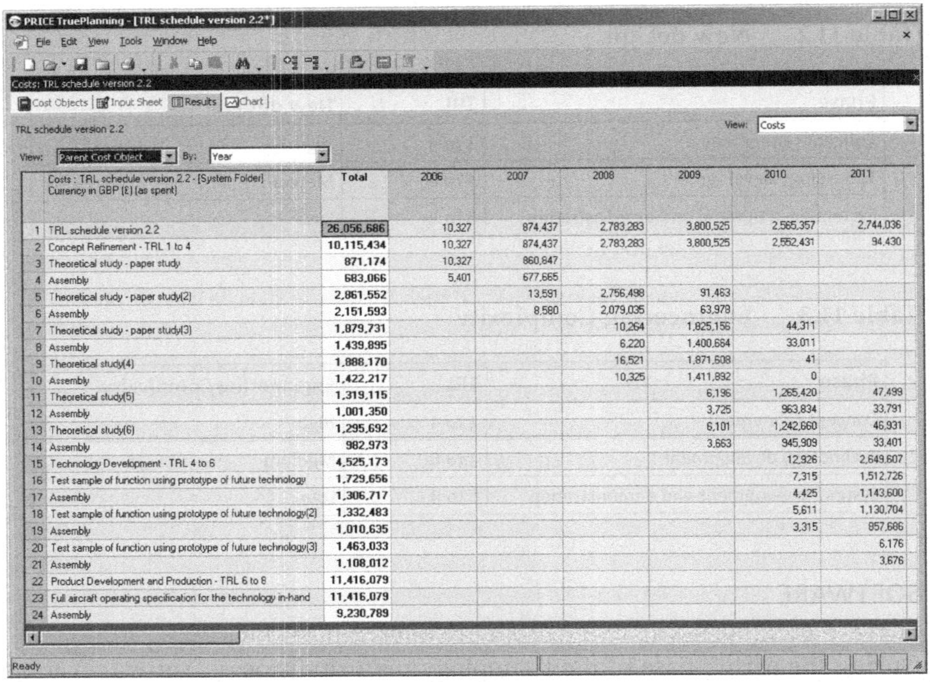

	Costs : TRL schedule version 2.2 - [System Folder] Currency in GBP (£) (as spent)	Total	2006	2007	2008	2009	2010	2011
1	TRL schedule version 2.2	26,056,686	10,327	874,437	2,783,283	3,800,525	2,565,357	2,744,036
2	Concept Refinement - TRL 1 to 4	10,115,434	10,327	874,437	2,783,283	3,800,525	2,552,431	94,430
3	Theoretical study - paper study	871,174	10,327	860,847				
4	Assembly	683,066	5,401	677,665				
5	Theoretical study - paper study(2)	2,861,552		13,591	2,756,498	91,463		
6	Assembly	2,151,593		8,580	2,079,035	63,978		
7	Theoretical study - paper study(3)	1,879,731			10,264	1,825,156	44,311	
8	Assembly	1,439,895			6,220	1,400,664	33,011	
9	Theoretical study(4)	1,888,170			16,521	1,871,608	41	
10	Assembly	1,422,217			10,325	1,411,892	0	
11	Theoretical study(5)	1,319,115				6,196	1,265,420	47,499
12	Assembly	1,001,350				3,725	963,834	33,791
13	Theoretical study(6)	1,295,692				6,101	1,242,660	46,931
14	Assembly	982,973				3,663	945,909	33,401
15	Technology Development - TRL 4 to 6	4,525,173					12,926	2,649,607
16	Test sample of function using prototype of future technology	1,729,656					7,315	1,512,726
17	Assembly	1,306,717					4,425	1,143,600
18	Test sample of function using prototype of future technology(2)	1,332,483					5,611	1,130,704
19	Assembly	1,010,635					3,315	857,686
20	Test sample of function using prototype of future technology(3)	1,463,033						6,176
21	Assembly	1,108,012						3,676
22	Product Development and Production - TRL 6 to 8	11,416,079						
23	Full aircraft operating specification for the technology in-hand	11,416,079						
24	Assembly	9,230,789						

Figure 11.10 Sample results for schedule demonstration

on the drawing board because nobody was willing to invest in bringing it to the market place. In this chapter we have learnt how to evaluate the cost and schedule of the most futuristic technologies, which will assist in providing a business case and funding.

Parametric cost models can be used to estimate technologies of low Technology Readiness Level including the cost of maturing the technology, investing for the technology and the manufacturing of the technology.

12

How to ... Assess Software

Software has become more difficult to ignore in programs. In our everyday lives the world around us involves more software than ever. It provides the ultimate flexibility in terms of providing a capability rather than a product. Product capability and functionality can be changed at the move of a mouse. However, the cost of software, relative to the hardware content, has increased dramatically and needs to be considered in all estimates for future programs.

The main problem with estimating software projects is determining the size or magnitude of the project. Parametric cost and schedule estimating models are helpful to an organization because they provide numerous approaches to evaluating the size of a software program. Once the size of a software project can be consistently determined, the transfer of data into estimated resources required is a relatively straightforward task as there is a strong correlation between size, language, functional complexity, organizational productivity and the resulting effort.

Software size can be quantified in various ways, but this chapter will highlight the most common approaches. These can be summarized as follows:

- Source Line of Code (SLOC);
- Function Point (FP);
- Predictive Object Point (POP);
- Use Case Conversion Point (UCCP);
- Commercial Off the Shelf (COTS) software sizing;
- Home-grown sizing methods.

However, regardless of the method used for sizing, it needs to be recognized that all software is not equal in terms of its influence on the parametric cost model (see Figure 12.1).

The best example of this is reused software, which, because it has been previously designed, coded and tested, has no immediate influence on the resources required in a program; it costs nothing. However, when combined with other new software, which does require total design, coding and testing, then there is an integration effect. Someone, somewhere needs to integrate the new and reused software and this requires resources. Hence, an accurate sizing of the reused software is needed.

Coming somewhere between new software and reused software is naturally adapted or modified software: software which exists, but whose design, code or test procedures need an element of change to satisfy the requirements of the new program. This element of change needs to be quantified in any cost model. A particular case of adaptation is deletion. In this case, rather than starting with code, some of which is deleted and replaced, the existing code is deleted and not replaced, it is simply removed.

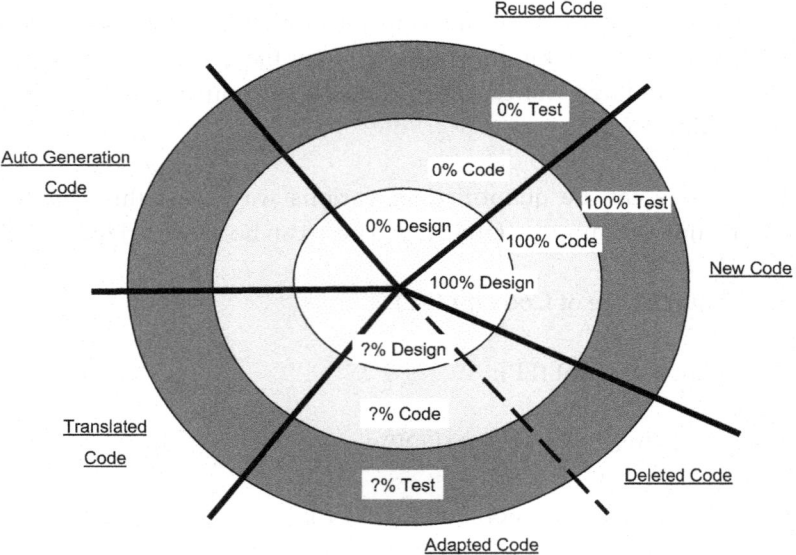

Figure 12.1 Software size

Normally in a cost model it is also reasonable to consider the translation of software from one language to another while maintaining the same design and function. Translation tools make this an attractive option when considering software development programs which replace existing systems. Finally, it is necessary to consider the amount of auto-generated software in your software program; code which is derived from model base design.

Source Lines of Code (SLOC)

The Source Lines of Code represents the number of logical lines of code of the software component. The first important issue with using Source Lines of Code (SLOC) counting is that the definition of a line of code can vary. A common parametric tool carries the following definitions of SLOC:

- may extend over more than one physical line;

- represents a complete statement in the implementation language;

- includes executable and non-executable statements, but not comments.

This value is usually determined by counting the number of delimiters (e.g. semi-colons).

The second question is when is a line of code not worth counting? The best examples of this are comment lines. Comments are generally excluded from the SLOC count, but the trick in estimating is to be consistent. Also, the non-executable statements (such as, software type declarations and data statements) are usually identified separately.

Why is this important? In terms of translating these lines of code counts into resource estimates, not all lines are equal. Comments, for example, are simple statements made in the code to provide an audit trail for the programmer. To produce the comments does not require a significant amount of effort. Producing the non-executable comments requires more effort, which can be determined when compared to the effort and resources required in producing the executable code.

The SLOC approach to estimating is historically favoured by contractors and suppliers for one very good reason, the SLOC can be validated and verified. At the end of the program is it possible to count them and refine the sizing methods.

Function Points

Source Lines of Code have consistently proven to be the most reliable sizing metric for estimating software development and support costs. However, it is difficult to determine SLOC for a conceptualized system without sound structured analysis of an extensive experience base. Consequently many software sizing methodologies have been invented and refined over the past quarter century to address the dilemma. The most popular technique, by far, is Function Point sizing.

In the mid 1970s, Allan J. Albrecht of IBM was tasked with measuring software productivity. He needed a method for measuring the size of a software development task independent of programming language and environment. In 1979, Albrecht published a paper[1] in which he first described a software sizing metric termed Function Points.

During the early 1980s, IBM and others employed metrics and Function Point counting procedures with some initial success. In 1983, Albrecht and John E. Gaffney, Jr. refined the Function Point technique by expanding function type definitions and counting procedures. When the economic usefulness of the Function Point metric became known, use of the metric expanded rapidly and quite spontaneously. By 1984, use of Function Points among IBM's client base had grown enough to form the nucleus of the present International Function Point Users Group (IFPUG).

WHAT ARE FUNCTION POINTS?

A Function Point is a unit of measurement that expresses the amount of functionality that a software system provides to a user. Function Points are an ISO recognized software metric employed to size a software system based on the functionality that is perceived by the user of the system, independent

---

1 A.J. Albrecht, 'Measuring Application Development Productivity', Proceedings of the Joint SHARE, GUIDE, and IBM Application Development Symposium, Monterey, CA, 14–17 October, IBM Corporation (1979), pp. 83–92.

of the technology used to implement the information system. Regardless of language, development method or hardware platform used, the number of Function Points for a system will remain constant.

The Function Point metric uses functional, logical entities – such as inputs, outputs and inquiries – that tend to relate more closely to the functions performed by the software as compared to other measures, such as lines of code. Function Point Analysis (FPA) is a structured technique of problem-solving. It is a method for breaking systems down into smaller components, so they can be better understood and analysed.

The Function Point generators found within many parametric cost estimating models are founded on the principles and guidelines *mandated* by the IFPUG.

FUNCTION POINT COUNTING STEPS

One of the initial design criteria for Function Points was to provide a mechanism that both software developers and users could use to define functional requirements. It was determined that the best way to gain an understanding of the users' needs was to approach their problem from the perspective of how they viewed the results. Therefore, one of the primary goals of FPA is to evaluate a system's capabilities from a user's point of view. To achieve this goal, the analysis is based upon the various ways users interact with systems.

Briefly, Function Point sizing involves counting five different categories of functions in a software application: inputs, outputs, inquiries, interfaces and internal files. These are functions the user of the application can see and identify. Each function is qualified in terms of complexity (low, average or high) and then multiplied by a corresponding complexity weight to achieve a Function Point count, a measure of the size of the function. The sum of Function Point counts for the functions is called the total Unadjusted Function Point Count (UFPC).

Next, 14 general system characteristics are rated with respect to degree of influence, zero through five with zero indicating no influence and five indicating an extremely strong influence. The characteristics include qualities such as data communications, performance and operational ease. The sum of the degrees of influence for the 14 characteristics is used to make a value adjustment to the UFPC. The result is the total Function Point count for the application.

Function Points are not favoured by contractors and suppliers as validation and verification is difficult even after the program has been completed. Two different function point counters could disagree on the precise number of function points that have been implemented.

Predictive Object Points

In recent years object-oriented technologies have become more popular in software engineering practice, both in terms of Object Orientated Design (OOD) and Object Orientated Programming (OOP). This growth has required software developers and their managers to rethink the way they have been estimating the size of their development projects. Traditional software measurement techniques have proven unsatisfactory for measuring productivity and predicting effort.

The SLOC metric and the Function Point metric were both conceived in an era when programming required dividing the solution space into data and procedures. This notion conflicts with the object-oriented philosophy. Traditional design techniques separate data and procedures while object-oriented designs combine them.

The Predictive Object Points (POPs) metric is one that incorporates the three dimensions of object-oriented software. Unlike traditional metrics, which are based on the data and procedure model of structured analysis, POPs are based on the objects and their characteristics. The POPs metric combines several measures to establish a metric suitable for predicting effort and tracking productivity.

The metric at the heart of the POPs calculation is Weighted Methods per Class (WMC). This metric examines each top-level class (or each distinct object from the user's perspective) and assigns a weight to the behaviours (methods) of that class. Once a value for WMC has been calculated, the POPs counter combines this with information about the groupings of objects into classes and the relationships between these classes of objects to assign the POPs count.

THE THREE DIMENSIONS OF OBJECT-ORIENTED SOFTWARE

The problem with traditional measures, as they apply to object-oriented solutions, is that they tend to measure one dimension of the software, the functionality. Without a measure of the complexity of object–object communications and the amount of reuse through inheritance, the functionality metric ignores crucial aspects of the software size. The functionality (behaviour of the objects) is an important piece of information when predicting effort, but to consider just this aspect could prove to be a mistake, particularly in a well-designed object-oriented solution. In addition to functionality, there is a level of complexity added to the software that depends on the amount of communication between the objects in the system. This complexity can add substantially to the size of the project. Increased inter-object communication requires more intensive design and test of the objects that are providing increased services to the system.

After functionality and complexity, the third dimension of object-oriented size is reuse through inheritance. Part of a good object-oriented analysis involves identifying groups of objects (actors) whose behaviours are similar enough to warrant them belonging to the same class or the same family of classes. A class is a description of the properties and behaviours that a particular object will have when instantiated. A group of objects with many similar behaviours is often designed as a base class (or parent class), having behaviours in common with several derived classes (or children classes), which inherit behaviour from the parent, add new behaviours that the parent does not provide and override those behaviours that require different functionality from the parents. Inheritance is a powerful feature of an object-oriented software system and has the potential for a substantial reduction of effort in certain software projects.

MEASURING ALL THREE DIMENSIONS

Predictive Object Points focus on the object-oriented analysis and design of a software system. They measure the important object-oriented aspects of that project – the classes developed, the behaviours of these classes and the effects of these behaviours on the rest of the system. They also incorporate measures of the breadth and depth of the intended class structure. The metrics included in a POPs count are:

> *Number of Top Level Classes (TLC)*: This metric is a count of the classes that are roots in the class diagram, from which all other classes are derived. Because the POPs count is intended for predictive purposes, it is important to be able to analyse the system with top level information. The Number of Top Level Classes, along with the Number of Children (NOC) and Depth of Inheritance Tree (DIT) provide a basis for an estimate

of the breadth and depth of the object-oriented system being described with this count. This adds the reuse through inheritance dimension to the POPs count.

Weighted Methods per Class (WMC): This metric is an average of the number of methods per class, where each method is weighted by a complexity based on the type of method, the number of properties the method affects and the number of services this method provides to the system. Methods can be divided into five classifications:

1. *Constructors* – methods which instantiate an object.

2. *Destructors* – methods which destroy an object.

3. *Modifiers* – methods which change the state of an object. A modifier will contain references to one or more of the properties of its own class or another class.

4. *Selectors* – methods that access the state of an object but make no changes to this state. These are the methods used to provide public access to the data that is encapsulated by the object.

5. *Iterators* – methods which access all parts of an object in a well-defined order. These may be used to visit each member in a collection of objects, performing the same operation on each member.

Weighted Methods per Class encompasses both the functionality and the inter-object communication in the POPs count.

Average Depth of Class in Hierarchy Tree (DIT): Each class described can be characterized as either a base class or a derived class. Those classes that are derived classes, fall somewhere in the class hierarchy other than the root. The DIT for a class indicates its depth in the inheritance tree, that is, the distance (in number of levels) from the root of the tree to that particular class. For example, in Figure 12.2, the DIT for Class C is 3 because there are three levels between the root, A, and Class C. The average DIT, along with TLC and NOC, is used to help establish the reuse through inheritance dimension and the overall system size.

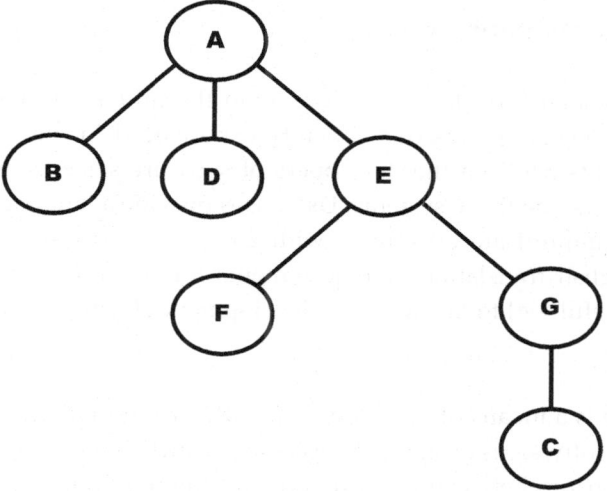

Figure 12.2 Sample inheritance tree

In this case the average DIT is calculated as:

$$\text{Average DIT} = \frac{(1 \times 0) + (3 \times 1) + (2 \times 2) + (1 \times 3)}{7} = 1.43$$

Average Number of Children per Class (NOC): Each class has zero or more direct descendants (derived classes). The NOC is a count of these derived classes. In Figure 12.2, the NOC for class A is 3 because it has three children – B, D and E. The average of the NOC across the parent classes is used to help determine the effects of reuse through inheritance and overall system size.

In this case the average NOC is calculated as:

$$\text{Average NOC} = \frac{(1 \times 3) + (1 \times 2) + (1 \times 1)}{3} = 2.0$$

Predictive Object Points, like Function Points, are not favoured by contractors and suppliers as validation and verification is difficult even after the program has been completed.

Use Case Conversion Points

Ivar Jacobson first introduced Use Cases[2] in the mid 1980s. One motivation behind their inception was the difficulty experienced in the communication of requirements between the developers of software systems and those who eventually consume those systems. Use Cases provide a language for software people to communicate effectively with users. A Use Case can be thought of as an English translation of requirements for software. They provide an extremely useful tool to facilitate the development of software that meets user expectations.

Use Case is a means of software sizing which is derived from considering the tasks the software is designed to execute. As such, it has a strong connection to the relationship between the software and the requirements. A project can consist of many Use Cases, each describing the tasks to be completed by the software, but like all software sizing techniques, that level of detail will be influenced by the stage of the program. Early in the program the level of definition will be higher than the final Use Case count.

Like Function Points, Use Cases are conceived from an external viewpoint. The Cost Engineer needs to consider the tasks to be satisfied by the software, rather than the code count. This technique also lends itself to sizing COTS systems as the inner working of the software is not required. Like FP and POP counting, Use Case counts are independent of the language in which they will ultimately be written. An example of a Use Case calculator is shown in Figure 12.3.

Each Use Case is given a title or name which describes the action required. For each Use Case, a complexity is assigned based on the number of transactions or behaviours. Each complexity is assigned a weight.

An actor is someone or something that acts on the system. It could be another system, a user, customer, administrator, and so on. The actor will be outside the system boundary being considered. Actor complexity is determined by examining how each actor is expected to interface with the software. Interface through a well-defined Application Programming Interface (API) will generally be simpler than an interface that involves reading streams of data. Certainly any interface that needs to accomplish intelligent communication with humans

---

2 Ivar Jacobson (1992), *Object-Oriented Software Engineering* (Reading, MA: Addison Wesley Professional).

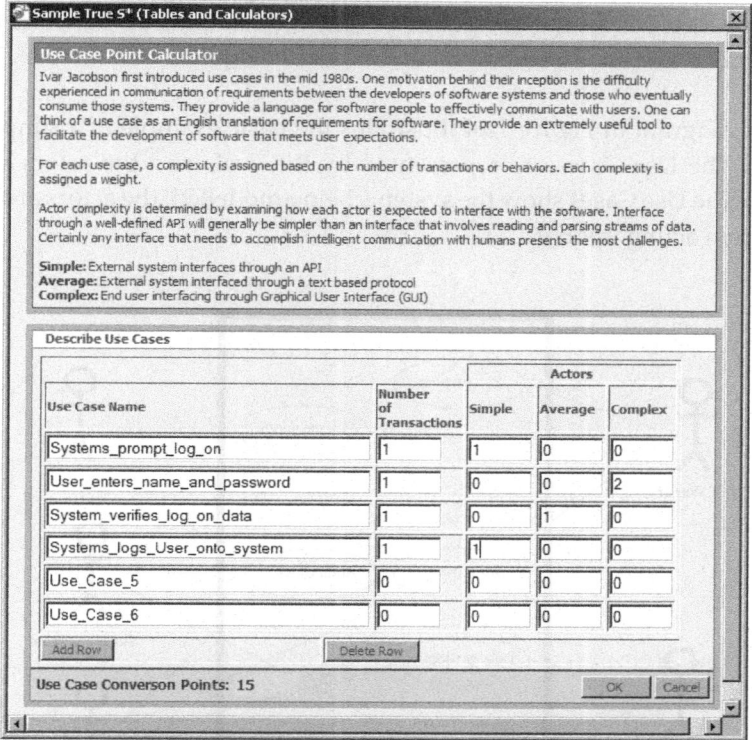

Figure 12.3 Use Case calculator

presents the most challenges. This is due to the random nature of responses from humans which are not experienced with systems communicating with other systems. Humans have a habit of poorly populating input fields – for example an account number with non-numeric characters – while another systems could be relied on to interface correctly.

In the example above the definitions could be:

- *simple* – external system interfaces through an Application Programming Interface (API);

- *average* – external system interfaced through a text based protocol;

- *complex* – end user interfacing through Graphical User Interface (GUI).

Using Unified Modelling Language (UML) the relationship between the Use Cases and actors are represented in a Use Case diagram.

The diagram in Figure 12.4 illustrates the basic example of a simple retail system. The Use Cases are represented by the ovals and the actors are stick figures. The Use Cases show the systems being modelled; the actors are outside the system acting upon it.

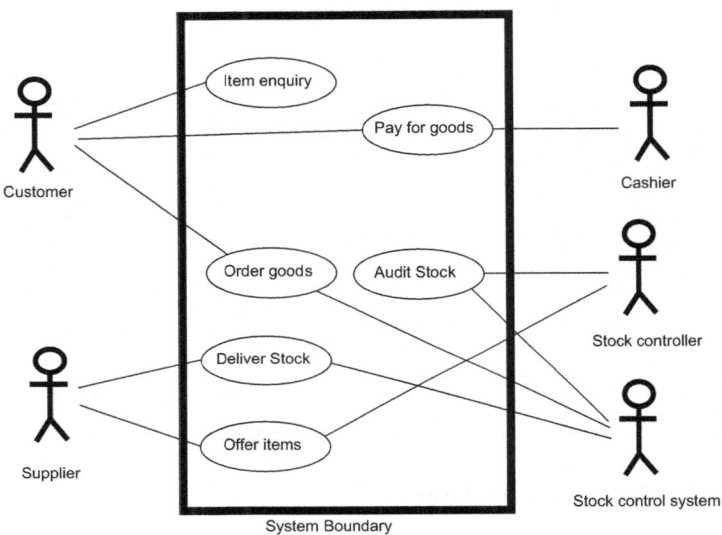

Figure 12.4   Simple retail system

Sizing COTS

One of the most difficult software components to size is Commercial Off the Shelf (COTS) software. This is due, in the majority of cases, to the source code not being available to the customer of the COTS supplier. So why bother sizing the system at all when you have a cost? Well, consider a program which is produced entirely from COTS software components. The developer has analysed the requirements, designed the perfect solution, evaluated the competition, acquired the ideal COTS components and only needs to integrate them to create the perfect, cheap system for the customer. But how can the magnitude of the integration activity be established without determining the size of the COTS components? This is the dilemma; the larger the components, the larger the integration task will become.

There are other sizing methods discussed in this chapter beyond SLOC counting. Function Points, Predictive Object Points, Use Case Conversion Points all lend themselves to the independent sizing of a COTS system because they take a view of the system from the users' standpoint and they are independent of the language.

Further, parametric cost models can solve the problem, including overcoming the concept of functional sizing. The size of a COTS component can be inferred from a rough description of its functionality.

Home-grown Sizing

Finally, there is estimating size by analogy. Many organizations store metrics of their current and historical projects to enable them to predict the size of future projects. Organizations will specialize in particular software types and will generate competitive advantage by reusing previous libraries of code and routines. Hence it is common to be able to size software by analogy. If the historical projects are analysed and the parameters are identified that have influenced the size of the project, then it is feasible to generate an organic sizing tool for a particular organization. The process for producing this type of sizing tool is similar to the process described later in this book for creating a cost model, except that the outcome is software size rather than cost.

Many parametric models are able to integrate with organic sizing methodologies or embed them within the model itself.

Summary

In this chapter you have learnt that size is a critical dimension when estimating the cost of software programs. Size has many definitions as not all code is simply new code; a good estimate should be considering at least the following software size categories:

- new code;
- reused code;

- adapted or modified code;

- deleted code;

- auto-generated code;

- translated code.

Although sizing a software program by analogy is extremely common, it does not lend itself easily to technical justification, traceability and audit. Other techniques for sizing code include:

- Source Line of Code calculators;

- Function Point calculators;

- Predictive Object Point calculators;

- Use Case Conversion Point calculators.

When commercial off-the-shelf code is being integrated into a software system, there are techniques for sizing the magnitude of this problem in parametric models.

13

How to ... Analyse Risk and Uncertainty

One problem with risk analysis is that risk means different things to different people. To discuss the subject we need to establish a common understanding, therefore we will consider risk as 'the probability of an event occurring and the impact of that event'.

If this is risk then what is uncertainty? Is there a difference and is it important? There is a difference between risk and uncertainty: the latter can be defined as 'a possible event, the probability and the consequences of which are unknown'.

The main difference is the predictive nature of the event. In the case of risk, a project has foresight of the event and can be proactive in respect to planning for it. In the case of uncertainty, a project has no knowledge of the event until it happens.

Problems are apparent in all projects. A practical engineering example is sending a satellite to a planet. The trajectory or path is calculated and a rocket sets the satellite on its course, but during the voyage the satellite has to tackle mid-course manoeuvres to ensure that it does not stray too far off course.

A project is similar. The project is scheduled and allocated a detailed financial budget to determine the most effective way of achieving the outcome. However, as the project progresses, money has to be spent to bring it back on course if problems caused by certain risks in its path are realized.

In terms of uncertainty, the scientists and engineers can design and build the best satellite in the world, only to find that totally unpredictable events occur which leave them without the capability they desired, for example: an explosion on launch caused by converting from 64-bit software to 16-bit (*Ariane 5*); an error in the focal length of the lens (Hubble Space Telescope); a satellite crashing into the planet due to conversions from metric to imperial measures (Mars Space Observer); and so on.

The UK Ministry of Defence has been using a Risk Process as a policy since the 1970s, when the Jordan, Lee and Cawsey report[1] identified the need.

To sell the idea of a Risk Process for a project it must have advantages. If an organization is going to invest time and resources in a process (by way of a project), there will need to be a pay back. The following list of reasons for performing a Risk Process is a good indication of its advantages:

- it enhances awareness of problems (risks) within the project – 'Manage the problems and the project will manage itself';

- it improves the reliability and competitive nature of project proposals or budgets;

- it provides budgets for the effective management of risks.

Listed below are other advantages that may be considered:

- it provides for a more realistic planning preparation in terms of cost and schedule;

- it provides an assessment of provisions which constitute a true reflection of risks;

- it facilitates risk taking (anticipation) through the enhanced knowledge of risks involved;

- it allows capitalizing on experience to improve confidence for future projects.

---

1   G. Jordan, I. Lee and G. Cawsey (1988), *Learning from Experience: A Report on the Arrangements for Managing Major Projects in the Procurement Executive* (London: HMSO).

Cause and Effect

The effects of risks are seen all too often. Projects sometimes experience problems as a result of assumptions in the program. These assumptions can start due to doubts or problems in the program. In the absence of facts, project teams tend to make assumptions in their place. Human nature is such that assumptions can be overly optimistic, particularly at the time of proposal, budgeting or approval, when there is pressure for a particular outcome.

Figure 13.1 shows some causes and effects of risks.

CAUSE	EFFECT
	• Overshooting costs/over consumption/cost overspend
Doubts Risks Problems Uncertainties	• Overshooting deadlines/ late deliveries
	• Inadequate performance/ non-conformities/wavers

Figure 13.1   Cause and effect diagram

If risks cause projects such difficulties, it would be beneficial if someone working on the projects were made responsible for their analysis. However, it is unrealistic for one person to be accountable for the risk; everyone in the project needs to be responsible for the identification of risk. But a clearly identified person should have the responsibility for the analysis.

The following are good reasons for performing risk analysis:

- to enhance the management's or customer's level of confidence in the cost figures;

- to help achieve a more realistic approach in the management of the project – proactive activity rather than reactive fire-fighting;

- to back up cost forecasts and to improve justification of costs;

- to ensure that the resources are provided that will enable the project to weather the storm when problems start to occur;

- to ensure that decisions are made on an equitable basis considering the total possible out-turn cost (nearest to the actual cost paid by the customer), not just the most likely cost.

Figure 13.2 shows how projects experience the effect of risks over their lifetime.

So at what stage of a project should we consider risks? At the start of a project a Risk Process can have an influence on the outcome of the project. With good management and forethought it is possible to avoid or mitigate risks proactively using a Risk Process. At this stage of the program the impact in terms of costs is relatively low.

As the program matures, the ability of risk management to influence the outcome of the program reduces. Conversely, the impact of risks increases as the program nears its end. As designs freeze, materials are committed, tooling is produced and the cost and schedule penalty increases if changes are required.

Take health as an analogy. Doctors will tell us it is better to exercise and eat healthily to prevent illness, than to have to take medicine to correct illness. In a similar way a project that is run on the basis of crisis management, with continual fire-fighting is like a body that needs surgery: very ill.

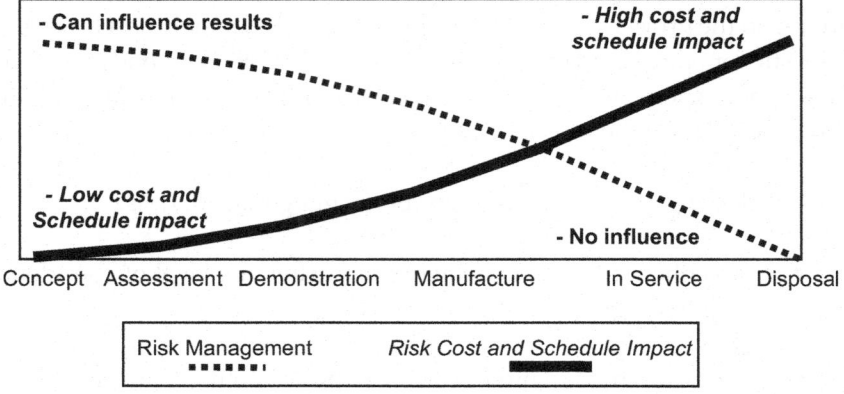

Figure 13.2 Consequences of risk

A Simple Risk Process

The steps in the Risk Process are given different names by different companies, consultants and Government departments. This depends mainly on the culture of the organization and what titles have been given to other project processes. The steps of the Risk Process generally cover the following activities:

- identifying all discrepancies which might arise during the project in relation to performance, cost and timing;

- taking all actions which enable the avoidance of discrepancies and/or the limiting of consequences;

- setting up relevant and adequate provisions to cover acceptable discrepancies;

- avoiding or limiting the consequences of the discrepancies which have been identified and guiding the project in such a way as to identify new discrepancies.

Figure 13.3 shows a typical Risk Process as a series of steps. The important point to note is the return arrow. The process is not a one-off exercise to enable the project to gain approval or a budget, but an integral part of the management of the project. As such the process needs to be performed or updated regularly.

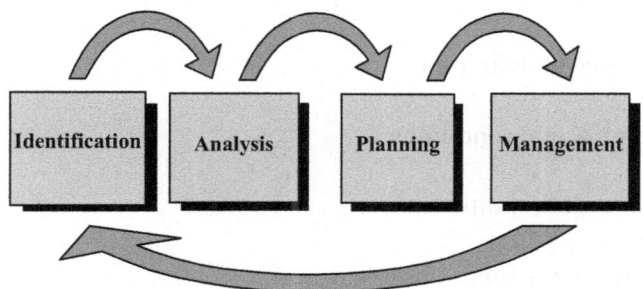

Figure 13.3 The risk process

Before any risk analysis can take place, a risk register is required. This may take the form of a project register, MS Access database, word processed document or spreadsheet. It will contain the risks that the project envisages will occur. Techniques used for risk register construction are given in Table 13.1.

Table 13.1 Risk identification techniques

Intuitive Technique	Advantages/Disadvantages
Expert	Relies on one or two persons representative of the experience of the entire organization The project team does not necessarily buy-in to the risks that have been identified for them
Questionnaires/Check lists	Ensure that all types of risks have been taken into consideration Can restrict the thought processes Not effective in simulating a risk aware environment
Interviews	Must be prepared in advance Technique depends on the interrogator or interviewer skills
Specialists/consultants	Acceptable in the short term Expensive Does not enable customer to develop internal know how and experience
Brain-storming	Clarifies assumptions and exclusions or misunderstanding Technique depends on the facilitator

A common technique is to use a Risk Prompt List. However, it should be treated as just that – a list – which prompts discussion and communication of problems or risks. It should not be used as a check list.

A prompt list should include risks that fall under the following main headings:

- technical concerns;
- estimating accuracy;
- degree of definition;
- contractual conditions;
- financial conditions;
- sub-contractors;
- personnel/corporate;
- schedule;
- policies;
- customer imposed risks.

Risk Quantification

Having identified the potential sources of risks a risk analysis cannot be performed until the project has completed a quantification of the risks. Following on from our earlier definition, risks are quantified on the basis of two primary dimensions: probability and impact.

PROBABILITY

Probability is expressed in terms of percentage and represents the likelihood of the risk occurring. A 50 per cent probability means that there is one chance in two that the risk will take place. If a risk has no probability it will be either: a certainty (100 per cent probable) and in the baseline program and budget; or an impossibility (0 per cent probable) and will never occur.

IMPACT

Impact represents the outcome or effect that the risk will have on the project. Impact is measured in terms of cost, schedule and performance. It represents the effect on the project if a risk occurs and is the difference between what the project anticipates in the future and what the project experiences in terms of out-turn cost, actual delivery date and product capability. Impact may be expressed in terms of: performance (qualitative or quantitative expression); cost, which can be expressed in terms of a *most likely*, *maximum* or *minimum* level based on the known degree of accuracy; timing, which, as in the case of cost, can be expressed as one of a range of values.

Impact valuation depends on the observer's position in time and in relation to the risk and actions which have been taken into account. Impact valuation must take into account normal working practice as well as the figures in the basic estimate. If it is generally accepted that scrap or rework is required, then these will already be included in the most likely cost figure. Therefore, to include them as a risk as well will only double account for their cost.

Impacts that can be surprisingly low in terms of their cost magnitude can nevertheless be labelled as high. For example, an 8 per cent increase in cost can eliminate all profit in some contracts so anything over 8 per cent is a high risk to a contractor. A late delivery or missing milestone can give a contractor serious cash flow problems, so a one-month slip can represent a high risk.

Qualitative risks are often represented diagrammatically as a Probability Impact Grid (or PIG). An example is shown in Figure 13.4.

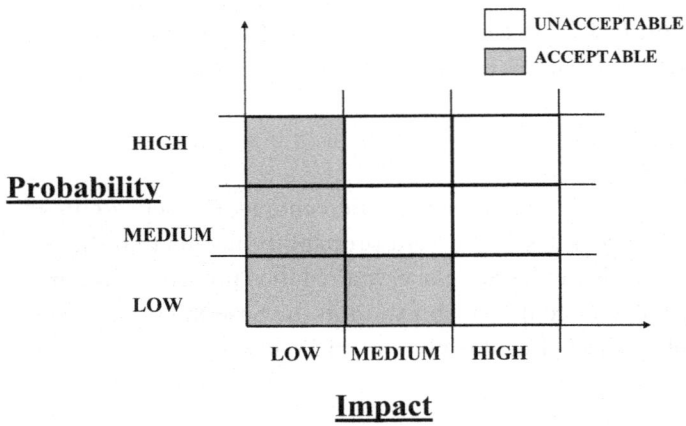

Figure 13.4   Probability Impact Grid (PIG)

RISK REGISTER

Every project being worked on should have a risk register, however, the form of the risk register can vary. Some will take the form of a simple MS-Word document, others will be sophisticated databases. Nevertheless, it is the fact that the project has a working risk register that is important, not the register's form. The essential features of a risk register, for the risk analysis, are:

- risk title;
- impact;
- probability.

The risk assessment is made before and after the risk mitigation actions have been implemented. It could well be informative to perform the impact risk analysis on both sets of data. After the analysis consider the difference between the results and the cost of the mitigation actions to determine if they represent value for money; in other words, if the mitigation actions are more expensive than just accepting the outcome of the risks.

RISK MITIGATION

Some people think of risk in terms of energy – it cannot be destroyed only changed into another form. In considering risk mitigation actions, there are two possible courses of action:

- alter the probability of the risk occurring;
- reduce the magnitude of the effect.

In this context the following are necessary: (a) anticipating the causes of risks and reducing their probabilities and, at the same time, (b) rectifying the possible effects of problems and decreasing their impact.

A project team can choose to:

- *eliminate* the cause – not assuming responsibility for the risks; using proven technologies;
- *transfer* the risk – appropriate in case of high probability risks; passing on the risk to a sub-contractor (specialist); sharing the risk with the contractor or with the sub-contractor (note: risk transfer will invariably come at a price);
- *avoid* the effect – appropriate in case of high-impact risks; duel source – acquiring the item from more than one contractor; keeping options open;

It should always be kept in mind that mitigation actions reduce the probability and gravity of, but do not necessarily eliminate risks. They do, however, allow the organization:

- to live with calculated risks;
- to reduce uncertainty.

These are the two main reasons for the risk analysis.

Risk Analysis

This risk methodology is intended to be used by Risk Analysts and project staff – not statisticians. The methodology provides a credible, practical and useable approach to produce understandable results. The methodology is simple, not sophisticated.

It should never be forgotten by the Risk Analyst that risk identification is more important than the risk analysis, which is why this chapter includes the Risk Process. As with all computer models, poor input data will result in poor results output. The objective is to allow the provision for risk to be optimized (neither too low nor too high).

RISK ANALYSIS METHODOLOGY

The methodology described here is based on a model which has two working levels and four steps (see Figure 13.5). The ground or lower level of risk analysis represents the overview or management level. During the steps which reside on this level the aim is to stand back and consider the project from a distance. The upper level is the detailed or technical level. The steps on this level are concerned with achieving analysis of the information.

In brief, the four steps are:

1. *Planning* – the methodology is discussed with the customer and agreed in the Master Data and Assumptions List (MDAL) or Cost Analysis Requirement Description (CARD).

2. *Cost Data Variability Analysis* – the cost forecast is examined for inaccuracy as a result of the data and cost forecasting model being used.

3. *Risk Analysis* – the risk register (including the inaccuracy of the cost model resulting from step 2) and uncertainties are combined.

4. *Reporting* – the customer is presented with the analysis in the format that was agreed with the customer and recorded in the planning step.

2. Data Variability Analysis	3. Probability/ Impact Risk Analysis
1. Planning	4. Result/ Reporting

Figure 13.5 The four steps of risk methodology

This process is ongoing throughout the life of the project and therefore quantification of risk is an iterative process.

Cost Data Variability Analysis

MONTE CARLO ANALYSIS

The majority of parametric models employ Data Variability Analysis, which may be accomplished at the system level (all Product Breakdown Structure (PBS) items), the assembly level (all elements indented below the assembly) and the element level (one PBS element). If Data Variability Analysis is performed at the system level, the model will make as many runs as specified through the PBS, randomly sampling any and all specified input distributions. Each pass through the PBS will result in a total cost estimate for the system. If 1,000 iterations are requested for a hardware or software system, the parametric model will make 1,000 separate system estimates.

An advantage of the parametric model is that it can perform Monte Carlo data variability analysis. This enables the user to enter a distribution instead of a single point for the inputs, and output a probability distribution curve. As a result, the output spans a range that will probably contain all the likely actual costs of the project. Thus the ability of predicting the outcome of the program cost is increased by the use of a range rather than a single point prediction

Parametric models have been designed to play a critical role in both cost and schedule analysis. The data variability analysis facilities in parametric models enable the translation of the uncertainty in the proposed hardware or software system's parameters into an assessment of cost and schedule risk. Variability can be run on the entire PBS or on a single element within the PBS. The data variability analysis process should include five basic steps:

1. structuring the PBS for the hardware and software system so that high risk elements can be identified;

2. determining the most uncertain input parameters for those elements;

3. quantifying the data variability for each parameter in each element chosen;

4. performing a simulation;

5. evaluating the results and iterating if required.

The alternative for parametric models systems is the FRisk[2] methodology for Data Variability Analysis. This method enables the user to input data variability around the input parameters as a three-point estimate as well as identify the correlation between components (see Figure 13.6).

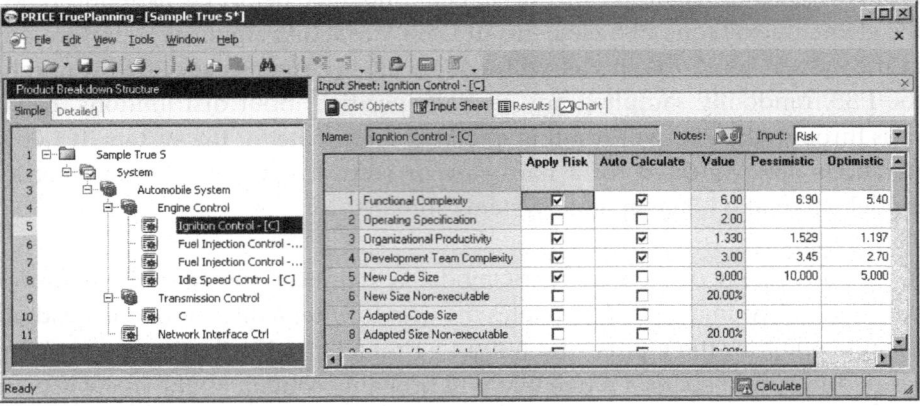

Figure 13.6   Example risk analysis inputs

---

2   P.H. Young, 'FRisk – Formal Risk Assessment of System Cost Estimates', AIAA 1992 Aerospace Design Conference, Irvine, CA, February 3–6, 1992.

Figure 13.7 demonstrates the ability to apply correlation, in this case to an information technology estimate, across PBS elements.

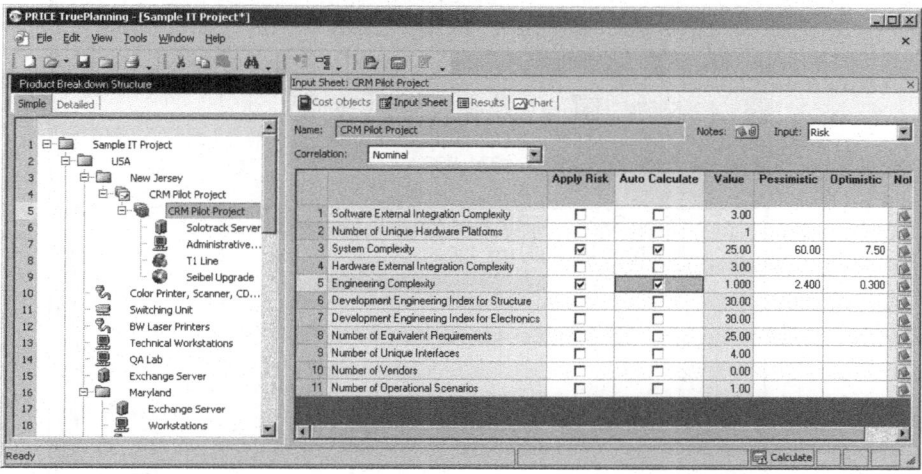

Figure 13.7   Risk analysis correlation across the PBS

The software then statistically compiles a probability distribution in the form of a 5 per cent confidence report (see Figure 13.8). As all the models are contained within the single framework, the risk technique is consistently applied across the models. Correlation of these inputs is also possible within these parametric models.

OPTIMISTIC AND PESSIMISTIC VALUES

Anyone who has conducted an analysis of this type will tell you the problem with this methodology is encouraging the engineers, who assess the data variability, to think broadly enough. When asked 'What is the likely maximum value?', most engineers will use their experience to consider an answer. However, if they can envisage the answer, it is a possibility and therefore not laying on the 0 per cent probability axis.

Figure 13.9 represents an automatic capability in the parametric model to ensure the probability distribution is broad. The model will calculate an optimistic and pessimistic value if required and, based on the phase of the project and the technology deployed, ensure that the first approximation of the probability distribution is wide enough or significantly pessimistic.

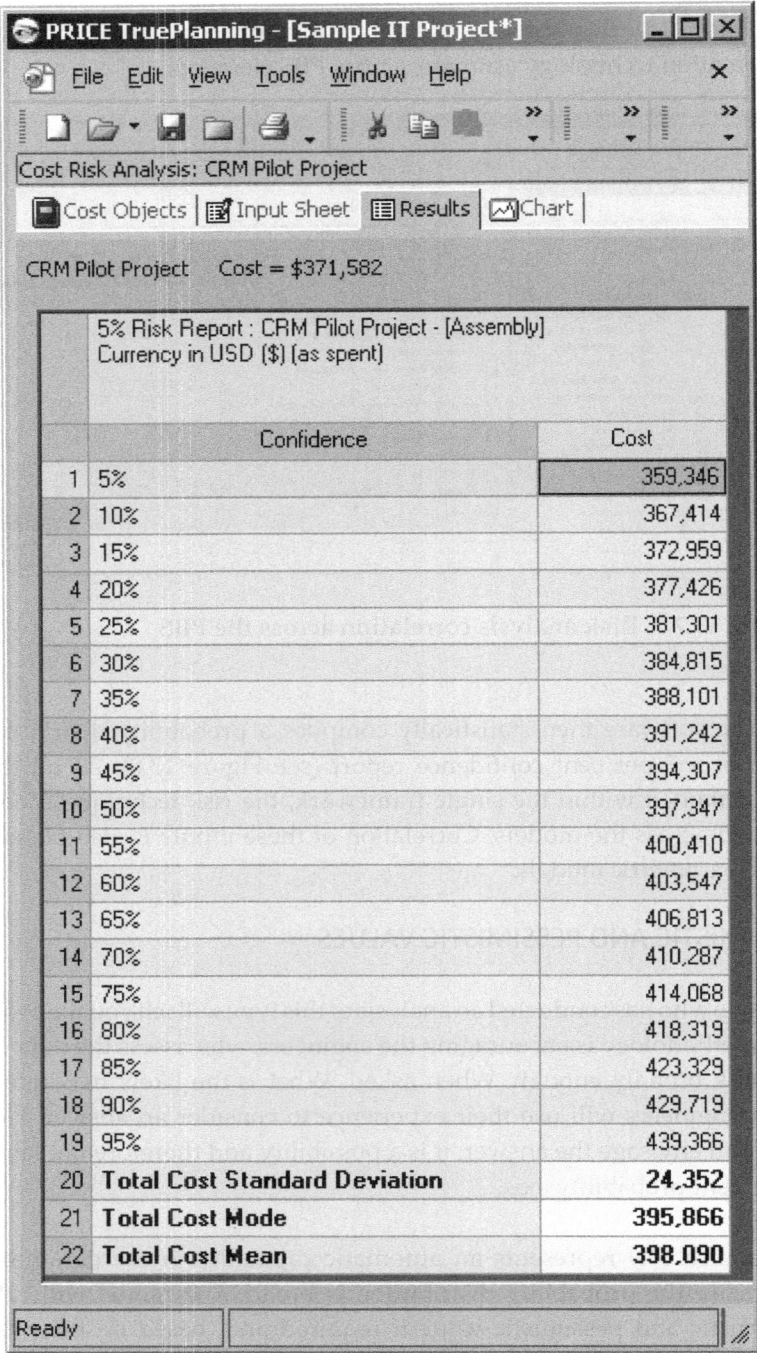

Figure 13.8 Risk output

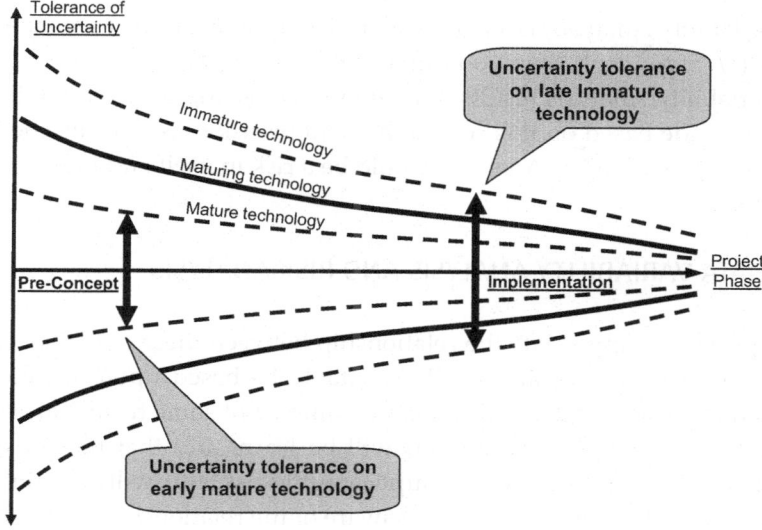

Figure 13.9 Defining optimistic and pessimistic values

BASELINE COST

A fundamental assumption of this approach is that the cost estimate will report the most likely cost figure as a baseline cost for the baseline program assuming normal working practices and no abnormal problems occurring. The risk analysis methodology will report the possible deviation from that baseline, the tolerance around the baseline estimate due to the data variability in the input parameters.

There is obviously a relationship between the risk register and the baseline program. If the program changes, because of the need to absorb mitigation plans, then the baseline program costs may increase and the risk analysis could reduce.

Parametric models have the ability to enter three-point estimates of their input variables. The models are then capable of performing a data variability analysis of these to predict the possible outcome in terms of cost or effort.

What they are not currently able to do is include the project specific risks reflected in the project risk register (for example political, social, economic and technological). The tolerance in the baseline estimate, generated by the

Data Variability Analysis, is a risk in all decisions made. If the tolerance is very narrow (+/- 5 per cent) then good financial decisions will be made, but if the Data Variability Analysis results in a large tolerance figure, then the financial decisions made based on this cost model will be risky. As a result, the output from the Data Variability Analysis should be a risk in itself and included in the risk register.

COST DATA VARIABILITY ANALYSIS AND RISK ANALYSIS

It is important to appreciate the relationship between these two analyses. The Cost Data Variability Analysis will inevitably be based upon historical cost information. To this end it will contain the impact of some historical risks that occurred. It is assumed that lessons will be learnt and that the project risk register that is analysed will not duplicate this risk, but will contain project specific risk. Risk Analysts need to be aware of the relationship, and as a result the person performing the Cost Data Variability Analysis is probably in the best position to perform the risk analysis.

The combination of the two analyses is illustrated in Figure 13.10.

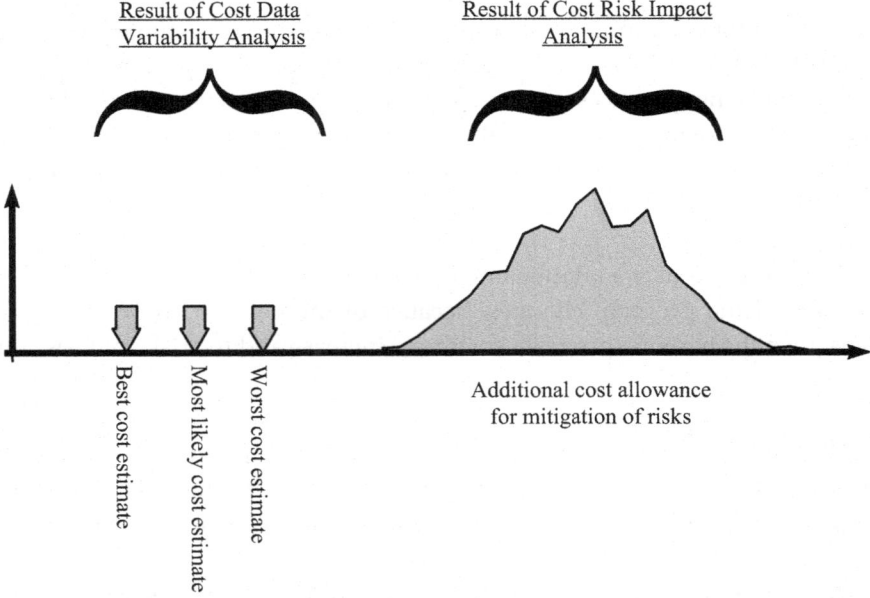

Figure 13.10 Results of analysis

Of most use to the project is the risk register ranked in descending order of importance or criticality. To achieve this, the results of the risk analysis are sorted on their *important index* column as shown in Table 13.2.

Table 13.2   Top risks

Top Risks	Impact			Prob.	Import. Index
	Min	ML	Max		
10 Procurement Strategy Inconsistent with Industrial Base	200	300	600	70%	33.90%
7 LTC Cuts due to Overspend by other Projects	150	250	600	75%	28.10%
Uncertainty / Residual risks	154.4	360	664	50%	25.90%
3 Standard of Living Conditions in 2015	200	300	400	40%	6.50%
1 Women in Crew	200	300	400	30%	4%
6 Defence Policy Changes	50	150	250	60%	0.90%
8 Tighter Environmental Legislation	50	150	250	90%	0.60%
Cost Data Variability Analysis	-178	0	220	100%	0.10%
9 Pressure to Reduce Operating Costs	0	100	200	75%	0%
11 Procurement Strategy does not Allow Latest Technology	0	100	200	50%	0%
53 Over-Conservative Design Margins	50	75	100	30%	0%
55 Requirement Growth	50	75	100	30%	0%

If the Risk Analyst is going to generate these results, it is essential that the project team trusts the process and realizes a return for their investment in time and effort. The table is an essential means for risk management to add value to the process and foster relationships with the project.

All information will be reported in the cost report and circulated for comment and retention. This will allow the comparison between the different procurement or design options on a cost basis, including the implications of risk, in each of the cases.

If you ask a professional Project Manager 'what is your highest project risk?' they will intuitively be able to tell you, together with their plans to mitigate it. If you then enquire 'what are your second and third priorities?' this might take a little longer to answer. Risk Analysis will enable full disclosure of the risk priorities to monitor through the importance index, thus feeding back valuable information to the staff that provided the raw risk register data to be analysed.

Summary

Programs rarely go to plan, and if they do they are rarely straight forward. For that reason risk analysis needs to be considered to account for the deviations and variations from the funding which is authorized or sanctioned. The probability and impact of events occurring in a program need to be assessed and quantified to ensure than an adequate, justifiable allowance is made to cope with these eventualities.

At the same time there is uncertainty in the input parameters which are used to populate parametric models when estimating. This is due greatly to the immature nature of the program (pre-design) and elements of technologies (TRL 1 to 4) which are being assessed at the time. Data variability analysis will determine the tolerance which results from the uncertainty and provide an indicator as to the risk in making decisions based upon the cost model.

When combined these analyses provide a comprehensive consideration of the funding requirements for a program allowing for all eventualities.

14

How to ... Influence Project Strategy

The acquisition and support of projects can vary in subtle or significant ways. Everybody in the supply chain is a potential customer and supplier. The way in which a system, sub-systems, assemblies or equipment are obtained can be modelled by a Cost Engineer and the disadvantages and advantages assessed. In this chapter we will consider how different strategies can be modelled and what to look for.

The first challenge is to create an independent 'should cost' of the components, which will enable an organization's procurement department to challenge the supplier quotations. Thereafter, the estimate needs to reflect the approach of the organization. This is more commonly the estimated cost of a systems integrator. Once the acquisition has been completed the cost of the support phase needs to be considered.

Acquisition Strategies

Increasingly, systems engineers responsible for software systems are relying on the incorporation of Commercial Off the Shelf (COTS) components to decrease the cost of and increase the cycle time for the delivery of new software systems. The same reasoning can be applied to creating hardware systems.

Intuitively, it seems obvious that software built and tested by someone else, that implements the functionality needed, would be cheaper and quicker than developing it yourself. It is important to understand, however, that COTS solutions are not nil-cost (and often not even low-cost) solutions. There are cost issues associated with the use of COTS solutions that need to be understood and managed in order to make sure that the COTS solutions chosen are the ones that will actually save time and money. All the activities associated with the development and sustaining of software systems that include COTS components, and the Cost Drivers that must be considered when attempting to make an assessment of the economic feasibility of COTS integration, need to be identified and estimated.

Regarding the COTS software it is assumed that:

- it is a commercially available software product – sold, leased or licensed;

- the source code is unavailable;

- there will be periodic releases with new features, upgrades for technology, and so on.

This is certainly not an all-inclusive description – there are many other types of products that might also be considered COTS, Military Off The Shelf (MOTS) being one example.

Activities Associated with Incorporating COTS Software and their Cost Drivers

Although in theory a COTS component should integrate into an application in much the same way that a component built in-house will integrate, there are issues associated with COTS products that are not present when all the components are custom made. In order to achieve a successful COTS integration, certain activities must be added to the project plan or (if already part of the project plan) must be expanded to incorporate COTS-related tasks. These tasks are briefly described with Cost Drivers outlined below.

EVALUATION AND SELECTION OF COTS SOLUTION

It is normally during the concept phase, or early in the requirement analysis phase, that the decision is made to investigate the feasibility of using a purchased rather than custom solution to deliver all or part of the functionality required. From this point, first a brief exploration should be undertaken to determine if COTS solutions exist, perhaps by searching the internet or visiting an appropriate trade show. This should be followed by a detailed analysis of the requirements for a solution which should include not only the technical requirements, but requirements associated with the stability and customer service record of the vendor as well as any cost, licensing or royalty considerations. Finally, a focused search of existing COTS solutions needs to be made to identify the one(s) that most closely meet the requirements. Depending on the number of COTS products available, this search may involve two stages – a quick search to identify the few most likely candidates, followed by a detailed evaluation of those few.

The primary Cost Drivers for this activity are the number of COTS solutions available, the number that will be focused on during the detailed evaluation, and an indication of the extent and rigidity of the requirements. The operating platform on which the solution is intended to run may also be a factor since quality and security requirements may limit the search.

PURCHASE (LEASE) OF THE SELECTED COTS SOLUTION

This is a relatively straightforward task in terms of identifying the Cost Driver. The purchase or licensing price will already be known. However, it is still important to include this in the analysis of a COTS solution because cost could have serious implications both up-front and throughout the life-cycle of the software solution.

INTEGRATION OF THE COTS SOLUTION INTO THE SOFTWARE SOLUTION

This is the heart of the COTS software integration and is one of the most poorly understood aspects of the process from a cost standpoint. This activity includes two specific tasks. The first task involves tailoring the COTS product to work in the system; there are certain activities that need to be completed in or around the software to get the COTS product working. Databases and other parameters need to be initialized; all or certain parts of the component need to be registered with the operating system; screens and reports need to be scripted; and other scripts may be required. These may not require the writing of any additional code, but they take time and they require effort to read manuals or attend training and then experiment with the actual tool to reach a level of competency. The costs of these activities are driven by the amount of functionality expected from the tool (number of reports and screens, number of Application Program Interfaces (APIs) called or number of parameters passed, number of scripts required, complexity of learning and using the product in terms of training available, quality of documentation, tools available, the presence of security and access control issues).

The second task involves developing glue code to interface with the COTS product. Glue code is the computer program that needs to be written to access the functionality of the COTS product and bridge any gaps between the data and formats of the organization's software and those required to communicate with the COTS products. The parameters that drive cost for this activity are the same as those that drive normal software development, that is, size of the functionality being included (Function Points, or number of APIs), complexity of the integration task (how complicated are the APIs), programmer skill and productivity, and the availability of tools.

It is important to understand that while the same factors drive the integration cost of a COTS component, the relationships between these drivers and cost will change somewhat because the integrators are dealing with unfamiliar code to which they have no access. It is also possible that glue code will have to be written specifically to deal with technical or performance problems in the COTS software – so there are additional Cost Drivers that indicate the level of technical or performance constraints in the COTS solution.

TESTING OF THE SYSTEM WITH COTS COMPONENTS INTEGRATED

Naturally, testing is not something that is required only because of the integration of COTS software. System level testing costs should be increased to account for the fact that external code is being integrated into the organization's system. The Cost Drivers for this increase in system integration and test costs are the amount of functionality being integrated and the required quality level of the finished product. Moreover, in a system integrating several COTS solutions, testers will sometimes discover (during the system tests) that incompatibilities between COTS solutions make it difficult (or impossible) for them to work together. Hence, if multiple COTS components are being integrated into a single system, this will also impact on the cost for the system test.

EVALUATING AND INCORPORATING UPGRADES DELIVERED DURING DEVELOPMENT

Depending on the stability of the product, the product vendor and the length of the development project, the possibility exists that the COTS solution will be upgraded while the integration effort is underway. In some cases upgrades will be required to incorporate necessary bug fixes, in others the upgrade will need to be evaluated and a decision made regarding whether replacement is warranted. If the decision is made to upgrade, the Cost Drivers can be determined thus:

- If no new functionality is being added – costs are driven by the amount of functionality already incorporated and a measure of vendor quality.

- If new functionality is being added with the upgrade – in addition to assessing replacement costs with the drivers above, efforts to tailor and write glue code based on the new functionality being added will need to be re-evaluated.

Activities Associated with Sustaining COTS Software and their Cost Drivers

Once a COTS solution has been deployed, the costs do not end. Just as the in-house developed components of the software need to be supported and sustained, so must the purchased ones, and, of course, there is cost associated with this. The following activities need to be accomplished during the deployment and life-cycle of the software.

MAINTAINING LICENCE AND/OR PAYING ROYALTIES

The licence fee is of course the Cost Driver for licensed COTS. If royalties are being paid, that cost is most probably also driven by the number of installations. Once again, this is a simple yet necessary part of any cost analysis of COTS solutions.

EVALUATION AND (POSSIBLY) INTEGRATION OF COTS UPGRADES

Just as upgrades had to be assessed during development, updated versions of the product will need to be evaluated and integrated. This is often necessary even in cases when no new functionality is required because upgrades also fix bugs and keep the COTS solution up to date with evolving technology. The same Cost Drivers as cited above apply.

INCORPORATING BUG FIXES ASSOCIATED WITH TAILORING, GLUE CODE, AND TO COMPENSATE FOR BUGS IN COTS COMPONENTS

COTS software, just like in-house code, is not defect free, and the cost to correct bugs related to the COTS software may differ from typical repair costs significantly because of the lack of available source code and a related lack of understanding of exactly what's going on inside it. Additionally, there may be a need to identify bugs in the COTS component that the vendor is unable or unwilling to fix and for which workarounds in the glue code will need to be developed. The Cost Drivers for COTS repair costs include a measure of the functionality incorporated, a measure of the COTS quality, the complexity of working with the COTS solution and a measure of the vendor's responsiveness.

Other Costs to Consider

There are other, often unconsidered, costs that may be associated with COTS integration that will need to be considered when performing a cost benefit analysis of a COTS solution.

COSTS TO KEEP COTS CODE IN ESCROW

Depending on the type of software system being built, there may be a requirement that a copy of the code for the COTS solution be held in escrow in case the vendor goes out of business or drops support of the COTS solution during the lifetime of the product. Some vendors charge for this service.

COSTS ASSOCIATED WITH REPLACING A COTS SOLUTION

Another situation that may arise involves repeating the entire process of evaluation and integration to replace the existing solution with one that has improved functionality, better quality or technical support or more closely meets current needs. This should be considered when evaluating the risks of a COTS solution.

There is much evidence to suggest that a well thought out, well implemented COTS solution will improve the speed and expense of a software development project. There is further evidence that these benefits are not as great as most of us expect them to be. It is certainly tempting to think, when planning a project, that creating the reports and charts will be a breeze because they will all be implemented with COTS products. Care needs to be taken when planning a project to consider all of the activities that should be performed and to schedule time and resources to complete these activities adequately. It is also important to look at the long-term issues associated with the solutions selected.

Multinational Influences

The globalization of business has created the opportunity to source equipment from beyond the normal national boundaries. This can provide opportunities to take advantage of nations with lower labour rates, lower raw material costs or higher productivity. A parametric model needs to be able to reflect this difference and provide a project with different scenarios for the most appropriate source nation of any given equipment.

Simply considering the exchange rate of the two nations is not enough. Figure 14.1 provides a good demonstration of the approach to be applied. Consider a specification for a suitcase with a given standard of material and quality. You could take this to a manufacturer in the United States and they might charge $100 for the product. If you were to take it to be manufactured in Greece, you would benefit from the lower labour and material costs. This can be calculated using a conversion factor or Purchasing Power Parity (PPP).[1] The Purchasing Power Parity is based on the cost of a basket of goods in different countries with the base country being the United States. A parametric model will convert the manufactured equipment using the PPP to reflect the

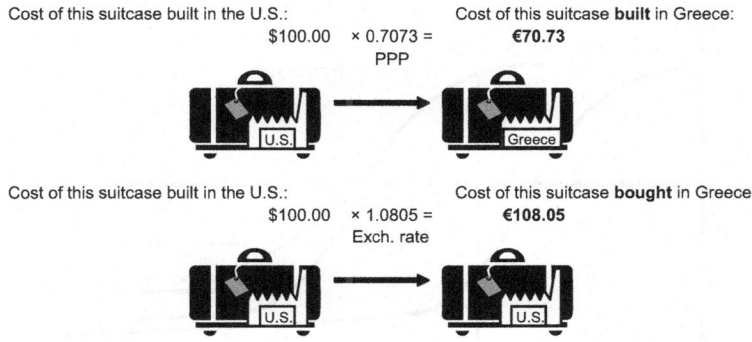

Figure 14.1 Multinational projects

---

1   http://www.oecd.org/std/prices-ppp.

productivity of a specific nation. Hence, for this suitcase the local cost in Greece would be €70.

The scenario would change if you were to take an American suitcase, built using American labour and materials, to Greece and sell it in Euros. In this example the price might be €108 and you would need to include the exchange rate in your calculations as you will ultimately need to pay the American employees for their time and American suppliers for their material in US dollars.

Program Influences

The effect of program decisions that are made or not made during a project's life can have profound cost implications. For example, a parametric model will include technology improvement algorithms. These will influence the cost of the procurement process based simply on the timing of the procurement.

Figure 14.2 demonstrates the influence of technology maturity on a project. Two technologies are plotted for a common operating environment or industry; a date has been established from cost research which indicates the technology maturity date. The maturity of lower complexity technologies will occur prior to the maturity of higher complexity technologies. Market forces will influence the cost of those technologies. Before the maturity date there will be very few suppliers of the materials and services required to deliver that technology, also few skilled staff available to provide the technology. Hence, a monopoly situation will encourage higher costs.

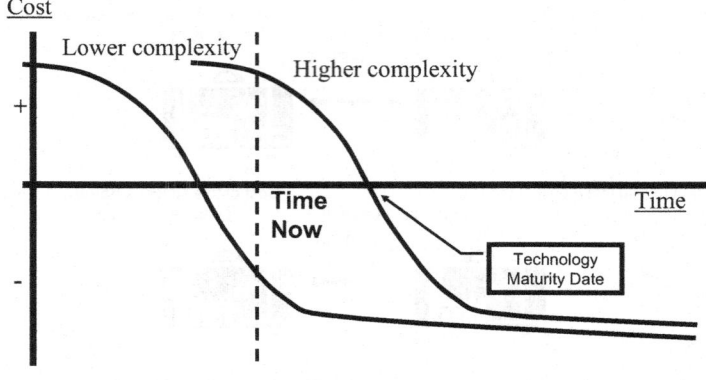

Figure 14.2 **Technology maturity related to acquisition**

As time passes, more suppliers enter the market place and competition will begin to drive down costs. Indeed there is a cost benefit in using mature technologies after their Technology Maturity Date.

Defence programs are constantly jumping to the next technology to ensure superiority in the battle space and maintenance of capability. This effect negates any savings, due to delays in program go ahead, resulting from technology maturity. The parametric models are unique in their ability to forecast future technology trends. It is possible to use the systems to extrapolate these trends to predict the likely cost of future technologies. This is seen as an important capability; although the configuration and implementation of that future technology is not yet known, parametric modelling does provide a traceable and justifiable method of predicting the future cost.

What influence does this have on acquisition strategies? By careful selection of the technologies used in a solution, the cost of the product may be reduced. Simply not making a decision may also result in seeing a reduction in the cost of projects. This is a risky strategy, however, as there are other influences that can be invoked when a decision is not made, but delivery dates are held firm.

As parametric models have the ability to estimate schedule as well as cost, they will do so on every occasion. Schedule is based on the type of technology to be delivered and the amount of that technology. From a top-down perspective, large amounts of very complex systems take longer to develop and produce than small amounts of very simple systems.

Having determined the optimum schedule for a system delivery it is now possible to establish the difference between the optimum schedule and any stipulated schedule. For example, the lack of a decision to proceed with a project acquisition is likely to result in the schedule being compressed if the delivery date is a fixed milestone.

Parametric models take a bird's-eye view of this scenario and will determine the cost implications of any non-optimum schedule. Consider Figure 14.3. This graph provides a representation of the schedule penalty algorithm. In this example, having determined that the optimum schedule is 12 months, based on the magnitude of the technology to be procured, there is a tolerance zone around that estimate. If, however, the schedule is shortened too much, the

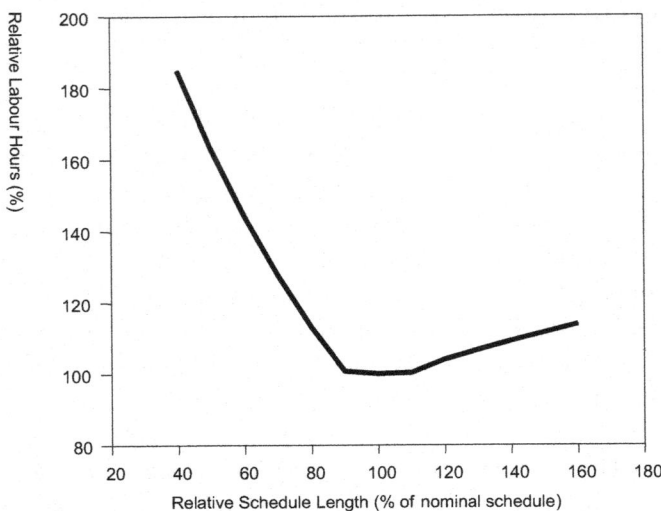

Figure 14.3 Schedule cost penalties

parametric model will determine a cost penalty to compensate for the fact that work will be going on in an inefficient environment. Work will be duplicated or wasted, simply because attempts will be made to work too fast.

The effect of stretching out the schedule is analogous, but not so severe, for similar reasons. As Parkinson's law[2] states: 'Work expands so as to fill the time available for its completion.' If the engineers have the opportunity to redesign the solution they will naturally attempt to improve it, beyond what is needed, simply because there is time, expending more time (and hence cost) in the process.

Project Phases

The other outcome of indecision is the effect of compression of the schedule phases, rather than the duration of the phases themselves. In other words, the development and production phases can still be optimum, thus avoiding any schedule cost penalties, but they are compressed overall.

---

2 Cyril Northcote Parkinson (1958), *Parkinson's Law: The Pursuit of Progress* (London: John Murray).

Consider the three acquisition strategies in Figure 14.4. The traditional approach is illustrated in strategy 1, when development is completed and production butts nicely up to the end of development. Strategy 2 represents a more expensive option: production is started prior to the end of development. The result is an increased workload for the engineers supporting the production team as they will be called upon to assist the manufacturing due to the incomplete nature of the design. Also there will be nugatory work conducted in the purchasing of long lead material and tooling which are ultimately not required when the final development prototype designs have been tested. When the development and production phases overlap, there is plenty of potential for labour effort and materials being wasted relative to the traditional approach in strategy 1.

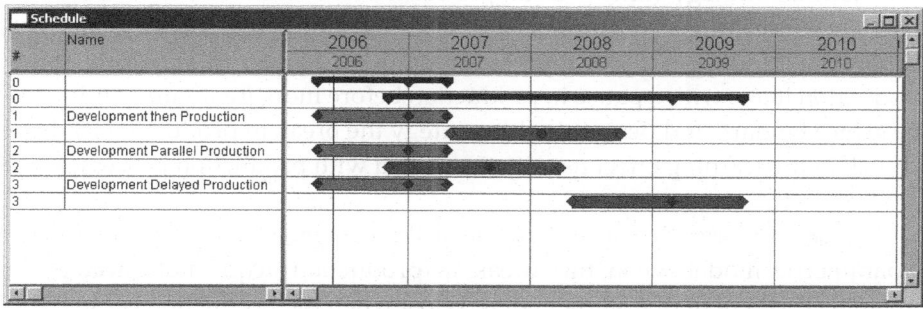

Figure 14.4   Relative relationship to program phases

The final strategy, 3, involves an extension in the gap between the development and production phases. When the time interval between development and production is longer than necessary, the parametric model assumes that time is used to value engineer the equipment item, in other words, to make it cheaper to manufacture. The resultant development costs reflect this task and they are higher than those for a typical schedule. The corresponding production costs take into account that value engineering preceded the production effort. These costs are therefore lower than they would be without the value engineering effort.

Lot or Batch Production

While considering the programming of acquisition strategies it would be wrong not to consider the influence of broken production. This is a subset of the scenario considered above – the traditional development followed by production. However, this scenario considers the effect of an acquisition strategy which breaks the production phase into batches or lots (see Figure 14.5). The first strategy shows the development of an item followed by the production of 100 items. The second strategy (lines 2 and 3) is based on the same principle, but the production phase is split into two continuous parts, each producing 50 items. Here the learning effect is continuous and the outcome similar to the first strategy.

The final strategy (lines 4 and 5) is sometimes considered by customers when there are budget constraints, but the implications are not always thought through. If funding is tight, then consideration is given to acquiring some production items with a pause in production before more items are acquired of the same standard and design. In this strategy the break in production leads to a break in the learning effect usually associated with production processes.

Figure 14.6 provides an indication of the cost effect that will be seen in a parametric model when this break in production acquisition strategy is pursued. The benefit of a learner curve is lost when the production ceases. The best that can be hoped for is that the workforce will *unlearn* at the same rate as they have learnt. Hence, for each month break in production, the parametric model goes one month back up the learning curve.

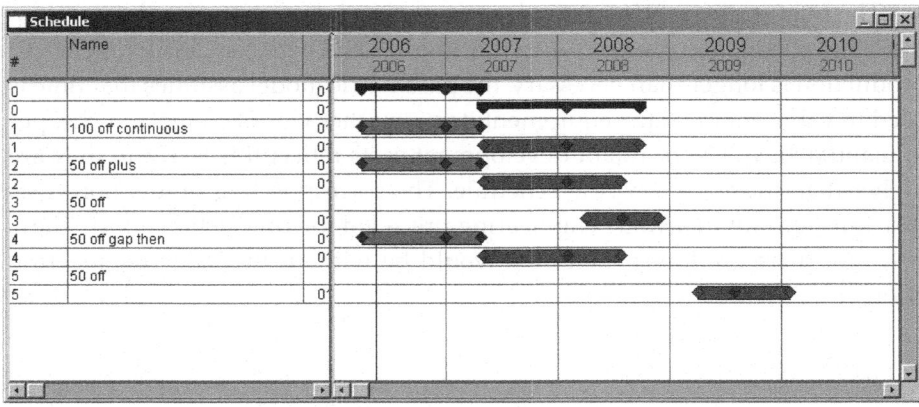

Figure 14.5 Continuous production or split production

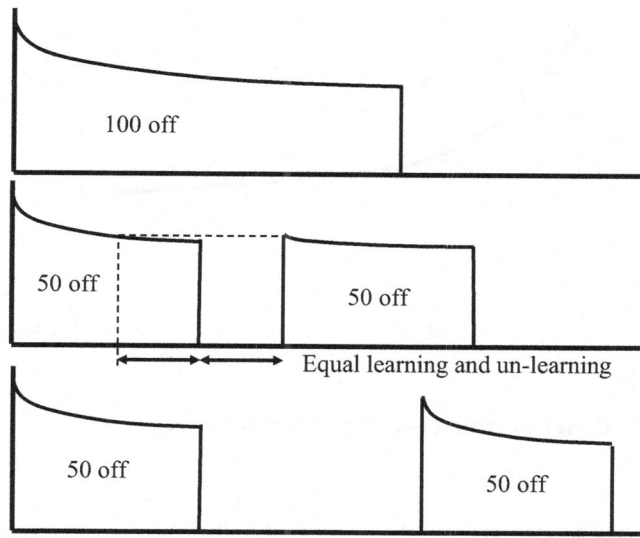

Figure 14.6 Batch quantities

Ultimately, in the bottom example, the break is too large to recover from and permanent loss of learning occurs; in any subsequent batch of production items, learning has to start again from the top of the curve.

This acquisition strategy differs from the next production program consideration; how a parametric model will allocate the cost of a large production batch across different programs.

BUYING FROM A LARGER PRODUCTION LOT OR BATCH

This capability is relevant to acquisition strategies which involve modularity. Technology insertion is made easier, and cost benefits can be realized, through the larger production quantities which result in modular construction. There will also be a reduced total development cost amortized over each item.

Figure 14.7 shows the traditional cost improvement curve or learner curve in many cost models including parametric models. The higher the quantity of items produced (100 off rather than 30 off) the lower the Unit Production Cost (UPC) (UPC_{100} is lower than UPC_{30}); parametric models share these curves, but retain the capability to allocate costs.

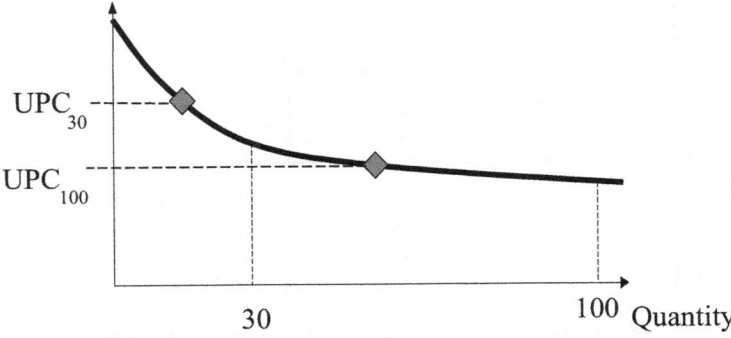

Figure 14.7 Benefits from cost improvement curves

As part of the procurement strategy, it will be advantageous to identify any equipment that can be common (or system-general), and therefore mass produced. For system-specific items (for example, chassis, hulls, wings, and so on), the quantity manufactured will be the same as the quantity required by that variant. For system-general items (for example, transmitters, receivers, software operating systems, and so forth), the benefits of quantity (see Figure 14.8 [for example 100 off]) will reduce the unit production cost (UPC) and the cost allocation will depend on the variant quantity for the system-specific items (for example 30 off at a UPC based on 100).

When considering the production quantity the Production Breakdown Structure is relevant (PBS). The level of detail of this PBS is not limited by the cost model, only the level of data that can be gathered at this stage. This means that the cost model can evolve with the project as more information becomes available in later studies or project phases.

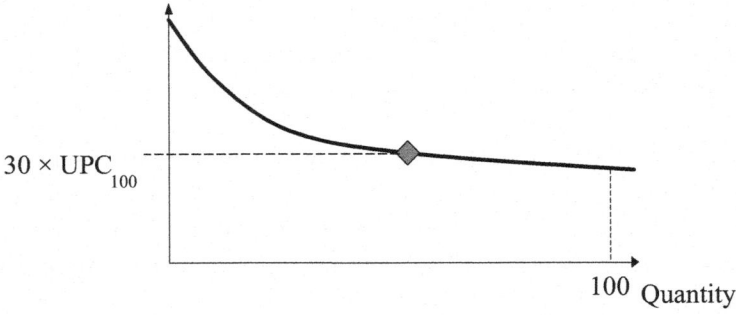

Figure 14.8 Benefits of the quantity allocated to the appropriate project

A Through Life Management Plan (TLMP) capability is also a feature of a good cost model (see Figure 14.9). As the system architecture becomes more stable through the program phases from concept to manufacture, organizations can request subcontractor bids for the equipment or components within their program. As the architecture design evolves, so must the estimating methodology, where the parametric elements can be substituted for purchased elements. Consequently the cost model can be used throughout the life of the project; neither this model will need to be changed, nor another employed.

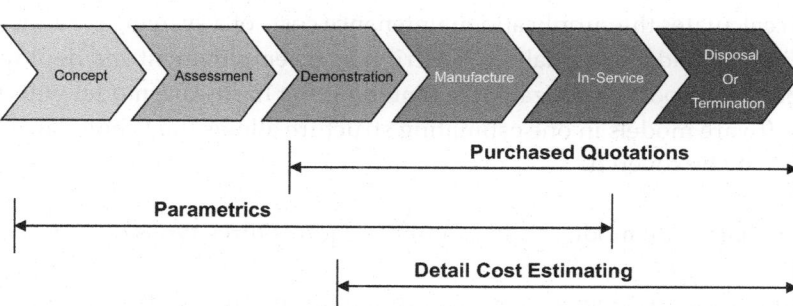

Figure 14.9   Life Cycle Cost capability of parametric models

Parametrics is ideally suited to acquisition strategy; it was specifically designed for use in the early conceptual phases of a project where little information on the project is available.

Parametrics also has the ability to predict the optimum schedule of a project using Schedule Estimating Relationships (SER) for both development and production. This will ensure that a realistic schedule is created at the outset, together with a realistic cost budget, to provide a sustainable Integrated Project Team (IPT). Members of IPTs realize that their careers and reputations are determined by the long-term viability of their projects. If the schedule and budget are unrealistic from the outset, there will be consequences.

Support Strategies

It is now widely recognized that support costs are equally as important as acquisition costs. The wrong support strategy can burden an organization for many years and can result in the lack of availability of a capability. Hence, the requirement is for a cost model that can determine the optimum support strategy based on cost and availability, considering both conventional in-house and Contractor Logistic Support (CLS) strategies.

Parametric models estimate the support cost of hardware, software and IT projects. The hardware model (Figure 14.10), for example, is a deterministic tool which calculates the supply and maintenance costs of a system, as a function of its reliability and maintainability description, its employment and deployment, and of the support organization. Combining the hardware model with the IT and software models in one estimating structure allows the Whole Life Cost of a project to be covered.

The hardware model, as an example, requires three types of data:

1. *Life-cycle inputs* – these are mainly hardware descriptors such as Mean Time Between Failures (MTBF), Mean Time to Repair (MTTR), cost and weight of spares, cost of test sets, Maintenance Concept, and so on. All these inputs can be calculated by the acquisition model, from acquisition descriptors such as weight, complexity, quantity, and so forth.

2. *Deployment inputs* – these are used to describe the repair and supply organization, the Operating and Support duration, and the number of deployed systems as well as their operating time.

3. *Global inputs* – these are of various types and include factors such as false failure rate, Turn Around Times (TATs), labour rates, or Safety Stock Coefficients, to try to obtain the best compromise (between cost and availability). The global inputs are set to defaults but can be customized.

The hardware parametric model generates various standard reports: the Detailed Cost Matrix, which shows the cost breakdown, test sets and maintenance workload; Initial and Replenishment Spares breakdown, availability and readiness; and the selected Maintenance Concept. These reports enable a trade-

Maintenance Concepts Diagram

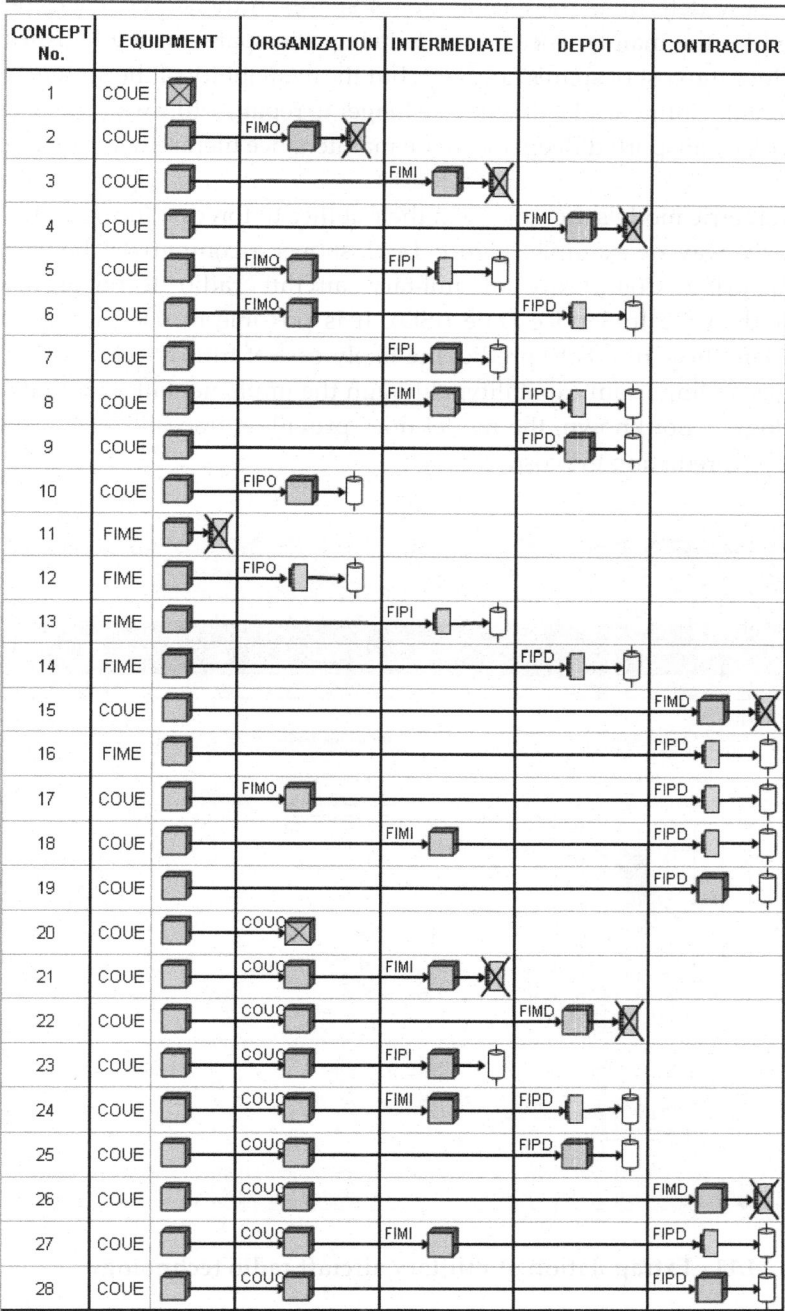

Figure 14.10 Possible hardware life-cycle maintenance concept

off in cost and availability using the Safety Stock Coefficients, to adjust the Fill Rates. These determine the availability of spares in the stores when they are required by the maintenance crew. Reducing the spares in the logistics loop will reduce the cost of spares holdings, but the availability of the system will be detrimentally influenced as it will take longer to repair, while waiting for spares to arrive or transported deeper into the maintenance hierarchy for repair.

Parametric models are unique in their ability to forecast future technology trends. By way of example, Figure 14.11 shows a comparison between the Manufacturing Complexity of military aircraft radio technology when compared to the In-Service Date (ISD). It is possible to use the systems to extrapolate these trends to predict the likely cost of future technologies. This is seen as an important capability; although the implementation of that future technology is not known, the model does provide a traceable and justifiable method of predicting the cost.

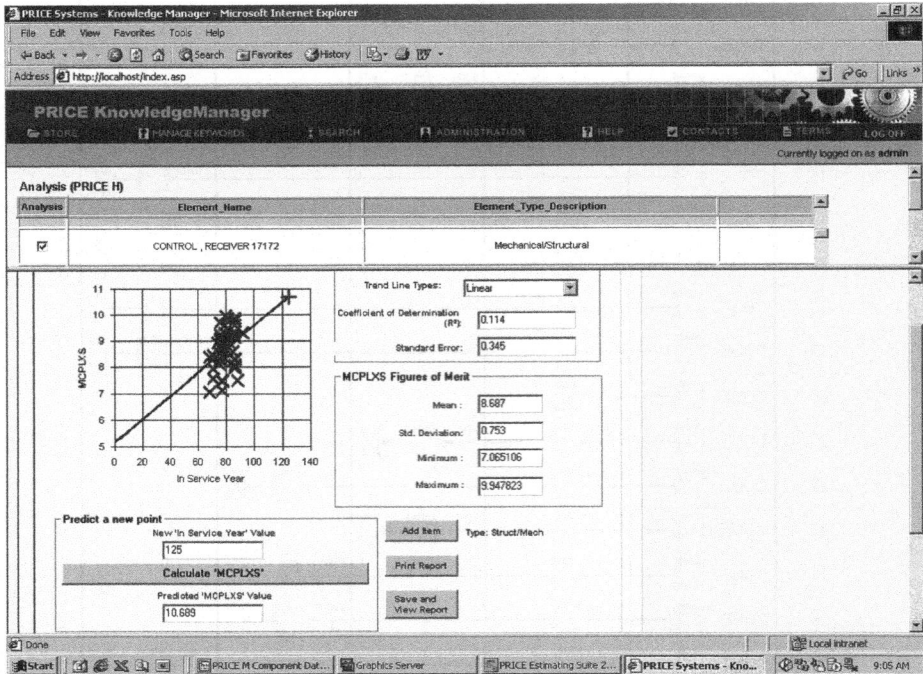

Figure 14.11 Extrapolation of military aircraft radio technology

Parametric models use a Product Breakdown Structure (PBS) – for this technology insertion application, the elements are arranged such that the *system specific* procurement items (airframe, chassis, hull), structure, electronics and core software are grouped. These are acquired for the full length of the program – 25, 30 years or more. The PBS will be used to define, in a consistent manner, the technical breakdown of the estimate.

The *system general* technology items which are likely to be acquired at insertion points in the program are grouped together for estimating purposes. For example, Figure 14.12 shows in one program structure:

- complete systems;

- current technology insertion;

- future technology insertion (10 years in the future);

- future technology insertion (20 years in the future).

The reason for this split is to identify the employment and deployment of the technology in the equipment architecture and their support needs.

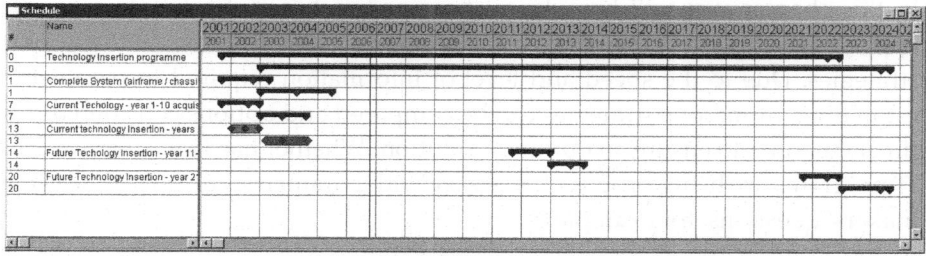

Figure 14.12 Top level technology insertion program

The result of the model is a time-phased expenditure which can be viewed as a table or graph (see Figure 14.13). Technology insertion or incremental acquisition is going to be the subject of the next chapter.

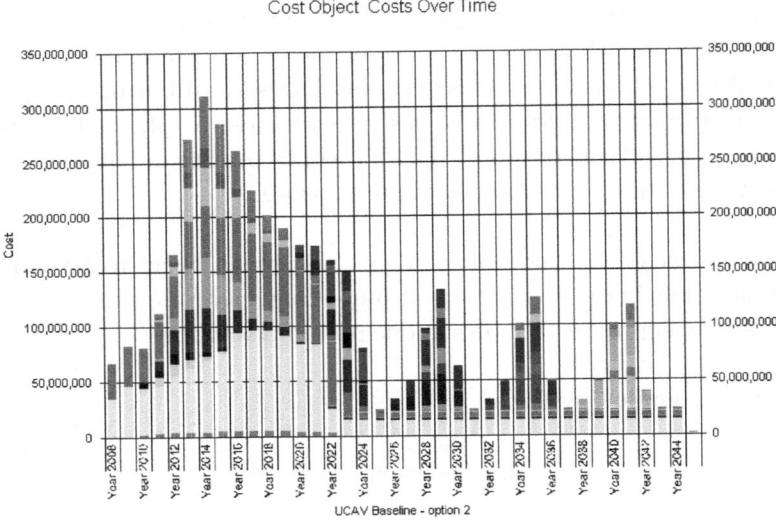

Figure 14.13 Time-phased expenditure

Summary

In this chapter you will have learnt how parametric cost estimating can be applied during the program's considerations of acquisition and support strategies. Due to the speed and level of information required in parametric estimating, numerous strategies can be compared and options considered in a timely manner to ensure an optimum solution. Whether hardware, software or IT programs are being proposed, it is easy with parametrics to answer decision makers' questions, for example:

- What if we buy this item, rather than make it?

- What if we source this item abroad?

- Will a change in technology save money or time?

- What is the optimum schedule for this item?

- Can we deliver this program in 2 years?

- If the schedule is compressed to the customer need, what is the likely cost penalty?

- If development and production are in parallel for a period, does it matter?

- If there is a gap in production between batches, will it affect the cost?

- This comes from a supplier making larger quantities than I need, has that affect been considered?

- If the architecture is modular, would that help?

- What is the optimum support strategy?

- Would contractorized support be cheaper?

- Should the program create a new intermediate support facility?

- Only part of the system will last all 30 years without upgrade, how should you model the rest?

15

How to ... Consider Technology Insertion

Unless engineers either are ILS managers, logistics specialists or have spent time in the support environment, it is common to use the terms upkeep, update and upgrade interchangeably. However, this leads to some confusion in projects and their funding.

It is generally recognized that the acquisition program represents the complete life of a system or equipment, from cradle to grave. This is usually initialled by a procurement process which is used to obtain the system or equipment, perhaps for the armed forces. This initial process is therefore termed the procurement program. It is likely that the procurement program will begin with a conceptual idea and then progress to an assessment and demonstration phase, before initiating full-scale expenditure on development and production of the system.

It is possible for there to be more than one procurement program within the total acquisition program (see Figure 15.1). The means of estimating the whole acquisition program is the topic of this chapter, including predicting the procurement programs.

In general, upkeep is the routine maintenance necessary to ensure that the system continues to function in the manner that it did when it was received. This includes scheduled maintenance (oil, paint, adjustment, and so on) and corrective maintenance (repair of broken engines or gearboxes with the same components). In the context of a home, this is the repair and maintenance that ensures our dwelling is clean and comfortable and keeps the roof over our heads.

It must be recognized, however, that as soon as a system is delivered, it starts to become less capable. In military language, the enemy adapts in terms

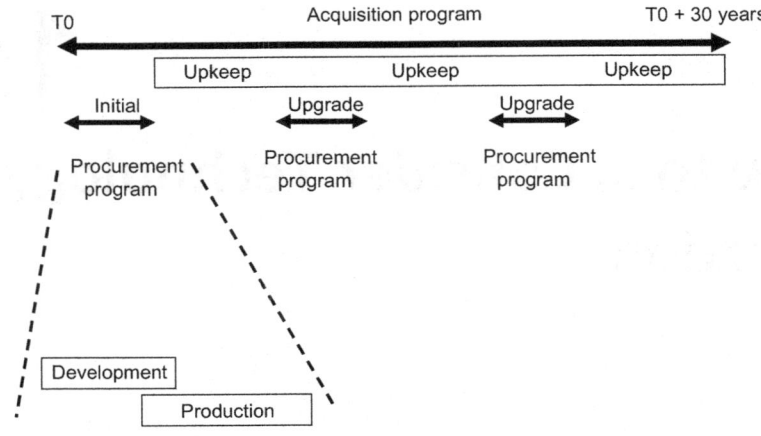

Figure 15.1   Acquisition and procurement definitions

of its mode of operation and doctrine to try to overcome the new capability that has been deployed. Maintaining the *current* systems will slowly lead to a reduction of the systems' effectiveness.

Update is the process of maintenance which ensures that the system is maintained in terms of both its ability to function and its operational capability. This is the means by which systems avoid obsolescence by updating the original technology to the latest version. If the electronics in a system need to be changed, why replace them with spares that contain obsolete chips and integrated circuits, when the option to replace them with current technology is available. When you repair the flat roof on your house, you don't just replace the bitumen and felt: you take the opportunity to consider new materials on the market just as you would consider new paint technology when decorating your rooms.

Upgrades have the effect of increasing the capability of the system or equipment during a maintenance event. The result will be a system or equipment that exceeds the capability effect of the item when it was originally delivered. These upgrades require the development of new subassemblies or subsystems that will be accommodated in the original equipment or system. As such they are likely to be the subject of a complete procurement program in their own right. Why replace one flat roof with another flat roof, when you could take the opportunity to upgrade to a pitched roof. When renovating our homes, why do we replace an old television set with a new flat-screen one? It's an opportunity to upgrade.

Why is the upgrade path interesting to defence procurement agencies? Because frequently projects are not adequately funded. Resources are tight and the capability that is desired is not always affordable. Upgrading means being able to procure the basic capability now with the opportunity of procuring the full capability at a later stage when the funds are available. Delaying procurement of the full capability, sometimes referred to as 'fitted for, but not with', can ease the cost profile of the project. For example, a ship or aircraft (the platform) can be purchased which has the *potential* to satisfy fully all the capabilities needed, but not all the systems are fitted at the In-Service Date or Initial Operating Capability date. This approach to platform procurement can be referred to as incremental procurement, as the capability will be acquired in increments.

Incremental procurement is by necessity planned; it is not an *ad hoc* process. History has seen many projects that have had a planned lifetime of 15 or 20 years extended because of funding constraints to 25 or 30 years. This is because mid-life upgrades have been needed to extend the life of a system or equipment through the inclusion of new subsystems and subassemblies. These are usually technically difficult and costly as, for example, when there is a need to cut and extend the fuselage, chassis or hull to accommodate the new technology.

The idea behind technology insertion is the creation of a system that is specifically intended to be upgradeable. The system or equipment, both the hardware and software, has been designed at the outset to be altered with future technologies with the minimum of effort. Such systems require characteristics that facilitate subsequent technology insertion, but this means in many cases that the system will be over-engineered for the immediate application.

This chapter considers the ability of Cost Engineers to estimate the Whole Life Cost of such acquisition philosophies and the cost modelling tool required to enable the exploration and justification of up-front investment. The reason for including this subject is that incremental acquisition or technology insertion has become a common topic amongst scientists and Systems Engineers, but to ensure that this approach to acquisition is not adopted in a casual manner, it is important that decision-makers are able to explore and justify up-front investment in system characteristics that will facilitate subsequent technology insertion. Hence this chapter describes the application of parametric models to the problem of technology insertion.

General Approach

While technologists and scientists can describe the issues or enablers that are needed for technology insertion to succeed, the list of topics that can be discussed, debated and concluded need to have cost influences related to them. Technologists and scientists have a good understanding of the issues and effects that are desired, but to enable the cost model to be influenced in the correct way, it is important that the cause is also properly understood.

The typical technology insertion issues to be examined when cost estimating are:

- support and maintaining Configuration Management information for multiple versions of the same system;

- discontinuity of labour resulting in peaks and troughs of workload;

- modularity of system architecture;

- procurement of systems with spare capacity in terms of:

    - space;
    - cooling;
    - bandwidth, and so on.

- documentation of system to facilitate change;

- utilization of openness systems which are defined as standard interfaces in the public domain, not requiring specification or investment. For example, ISO standards, RS232, Bluetooth, USB connections. There are multiple sources of suppliers. These interfaces can be referenced to; they do not require a specification;

- maintaining over an extended period the availability of relevant knowledge, skills and tools;

- obsolescence.

Technology Insertion Case Study

An Unmanned Air Vehicle (UAV) is the selected system for this case study adapted from a study with QinetiQ in the UK. To demonstrate how the options of non-technology insertion and technology insertion can be considered and compared on an equal basis using parametric estimating.

The two options to be estimated when considering technology insertion are shown in Figure 15.2. Option 0 (baseline) and Option 1 (technology insertion) are considered to be the worst-case situations and the extremes to model.

The options are described as:

> *Option 0 – 'Traditional' – Baseline UAV Design* – An option to consider the traditional method of dealing with technology in what is perceived to be a typical defence project in terms of a contracting and engineering solution. This is the baseline UAV design.

> *Option 1 – 'High Receptivity – Technology Insertion'* – An option to build new systems in an incremental manner with the philosophy of technology insertion planned from the inception of the project; hence it will have built-in receptiveness to smaller upgrades (technology insertion activities) through the life of the program. This is the adaptable UAV design.

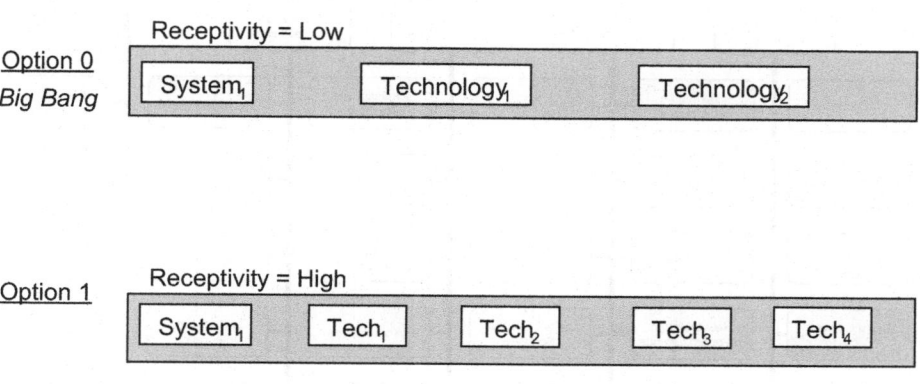

Figure 15.2   UAV options

DATA AND ASSUMPTIONS

A Product Breakdown Structure (PBS) can be agreed for each of the options to be parametrically modelled. For Options 0 and 1 the technologies were identical, but for Option 0 none of the items were conceived with the idea of subsequent technology insertion. There would be no technology insertion candidates.

Figure 15.3 shows the PBS that was considered.

Technical information and Cost Driver data were gathered against the elements of the PBS from the most authoritative sources. These technical inputs to the cost model determined the differences between Options 0 and 1. The algorithms within the cost model use these technical inputs to determine the appropriate costs to be generated.

In Figure 15.4 the degree of technology replaced or upgraded is indicated at each technology insertion point. Being less receptive to technology insertion, the systems of Option 0 would be more intense and, in the case of the propulsion and sensor systems, delayed until absolutely necessary.

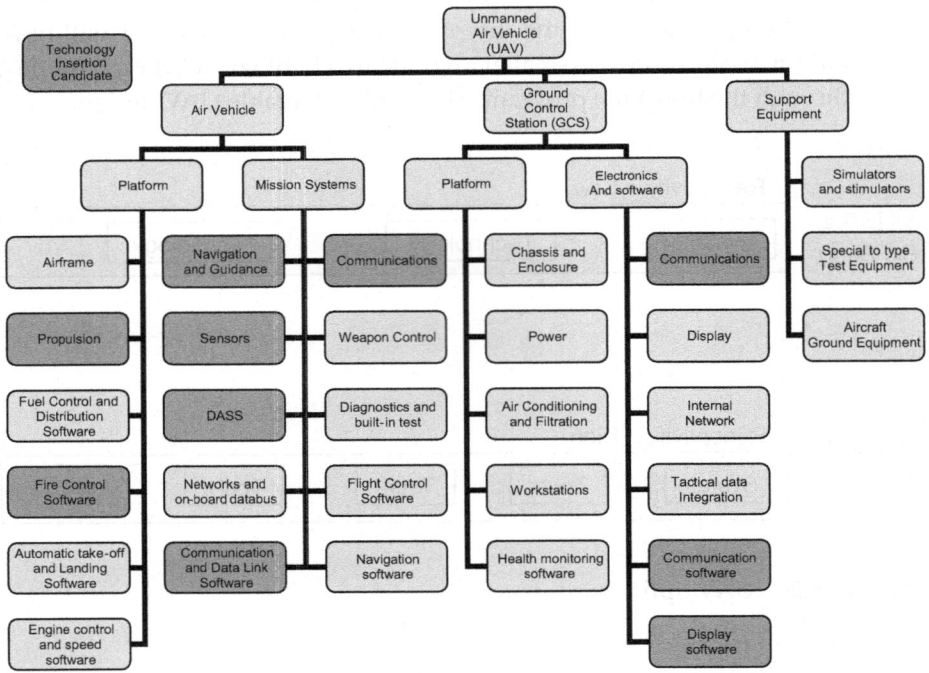

Figure 15.3 Option 1 – hardware and software PBS

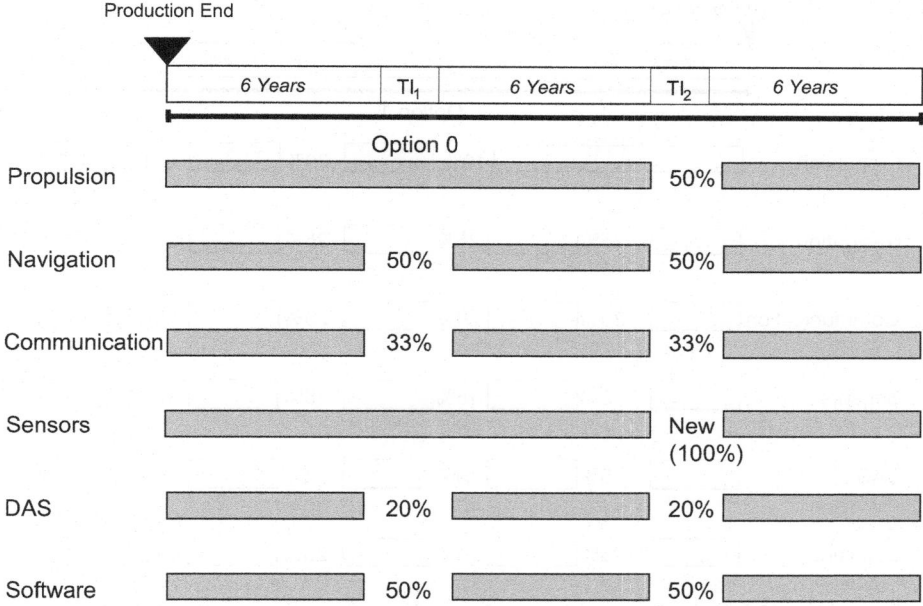

Figure 15.4 Option 0 – degrees of technology insertion

In Figure 15.5 the degree of technology replaced or upgraded in Option 1 is indicated at each technology insertion point. In this instance the opportunities to conduct technology insertion are more frequent and less dramatic for the system. Hence, it is considered that the capability would be continuously topped up during the life of the system with smaller, frequent technology insertion.

MODELLING CONSIDERATIONS

To model the technology insertion issues in a parametric cost model is relatively simple. Some of these issues are considered here.

Spare capacity

The baseline system would design the enclosures of the systems to house exactly the internal mechanisms and electronics. No consideration would be given to providing excess space for future upgrades. However, Option 1, which is designed for technology insertion at some point in the future, would be designed with excess volume. As the hardware parametric model is weight

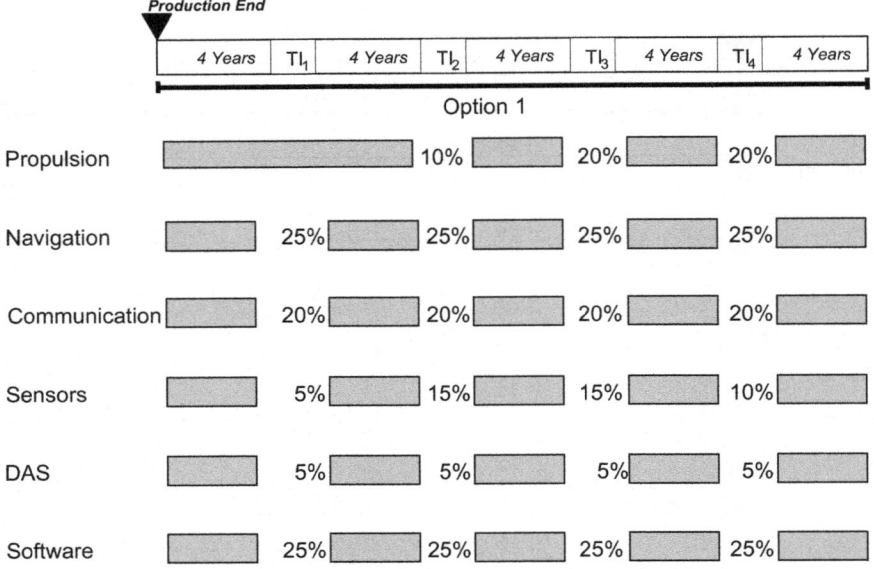

Figure 15.5 Option 1 – degrees of technology insertion

based and the structural and electronic weight is assessed separately, it is easy to indicate to the cost model that although the two systems have the same capability at the outset, the option considering technology insertion (Option 1) is likely to have more structural weight while both have the same electronics weight.

Spare capacity and modular architecture are two technology insertion issues that are summarized in the approaches drawn in Figure 15.6. On the left is a traditional approach, non-receptive to technology insertion. The architecture is a 'spaghetti' of connections and interfaces. The equipment is designed to fully satisfy the requirements of the current system, with no thought about future growth.

The architecture on the right has a common interface box. All equipment is connected via this interface: consequently, any future technology insertion only requires that the new technology interface with the one box, which will deal with all timing, compatibility and interface issues of the system. The architecture on the right is also over-sized. The weight of electronics is equal to the required initial need, but the structure has been overstated to ensure that there is space for electronics at any future technology insertion point.

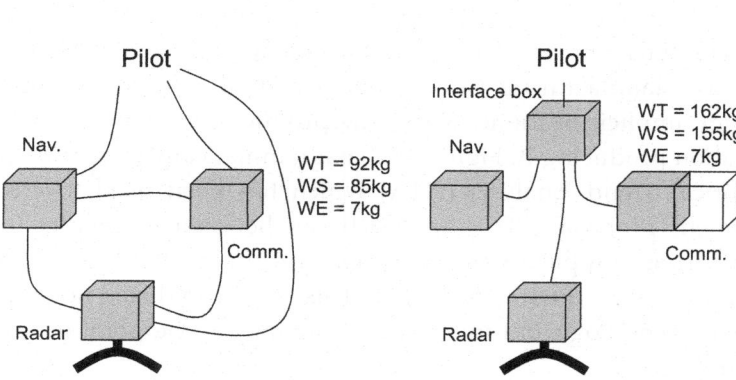

Figure 15.6 Possible technical architecture of receptive and non-receptive systems

Integration

Modularity of a system will lead to much simpler later integration activities. While it is possible that both options are simple to integrate at the production phase of the acquisition cycle, it is at the point of the technology insertion that the benefits of modularity will be realized. The baseline option (Option 0), never envisaging the possibility of having new technologies introduced, will be difficult to upgrade, while upgrading the technology insertion option (Option 1) will be simple.

Engineering

The cost of the initial design and development of the baseline option (Option 0) will be less than the technology insertion option. The baseline option will have Cost Drivers based on a new design and nominal engineering, while Option 1, with technology insertion capability, will need to have Cost Drivers based upon a new product and new technology increasing the scope of the design work. As they approach the technology insertion points, the Cost Drivers will be reversed as the technology insertion option (1) will require only a simple modification to an existing system (a modification which was pre-planned), while the baseline option (0) will need extensive modification to a design which was never intended to be modified.

Technology prediction

Input parameters in parametric models include normalized cost density defined as Manufacturing Complexity, which is a representation of the technology implicit in the item and the productivity of the manufacturing organization producing it. Figure 15.7 diagrammatically shows the principle of Historical Trend Analysis (HTA) which has been used to cross-check procurement proposals. This approach can be used to predict the likely Manufacturing Complexity of equipments resulting from future technology insertion based on time trend analysis. This can be predicted not only for the immediate technology, but for future technology at the point of technology insertion.

It is possible to plot Manufacturing Complexity against performance like, power (kw) and heat (joules). In Option 1 spare capacity is planned initially for power provision and cooling such that at a technology insertion point, these are already adequate and can accommodate new upgraded systems and electronics.

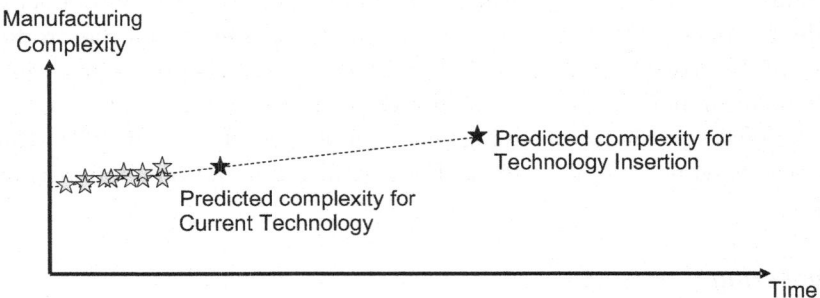

Figure 15.7   Historical Trend Analysis

Design for reuse

The amount of design effort that needs to be performed also varies between the two options. Starting with the baseline and describing this as nominal design effort in the parametric model, as an unexpected mid-life upgrade is approached, this can only be described as high design effort to accommodate the new technology in the existing platform. Conversely, on the Option 1 platform the initial design effort is increased to ensure that the system is built to accommodate new technology easily at a later stage in its life. This is particularly true for the software, which needs to be designed for reuse at the initial stages of procurement. This indication in the parametric cost model will result in higher design costs, but the payback will occur at the technology insertion points when the software is changed to accommodate new technology.

The software and hardware development teams will have issues regarding the continuity of the team. Keeping the initial development team's knowledge from Option 0, when no follow-on work is anticipated, is difficult. At least in Option 1 a technology insertion is planned, so the development team can see a reason to continue with an organization, as there will be new technologies to investigate, new knowledge to gain and personal professional development.

Maintenance

When it is necessary to model the cost of the whole life-cycle, then parametric inputs like the Mean Time to Repair (MTTR) are considered. Options 0 and 1 would be comparable in terms of MTTR, but the fact that the technology integration option is designed with modular construction and open architecture in mind will mean that when a technology insertion point is reached, Option 1 will reap a benefit in terms of the lower repair labour times required. The cost to disassemble the hardware or software and then integrate and test it again, incorporating the new technology, must be easier. This should apply equally for on-equipment upkeep as well as off-equipment upgrades and updates.

CASE STUDY RESULTS

The generated costs from the parametric models need to be peer reviewed to consider them for anomalies, for example typing errors. The outputs of the models are cost profiles and associated schedules. The appropriate discount rate can be used to generate a discount factor for each year of the project life. The total figures obtained from the WLC will be multiplied by the annual discount rate to obtain the discounted annual cost. These will be aggregated to generate the Present Value (PV).

Figure 15.8 clearly shows the initial increased investment in Option 1 when compared to Option 0, followed by the reduced procurement activity at the subsequent technology insertion points.

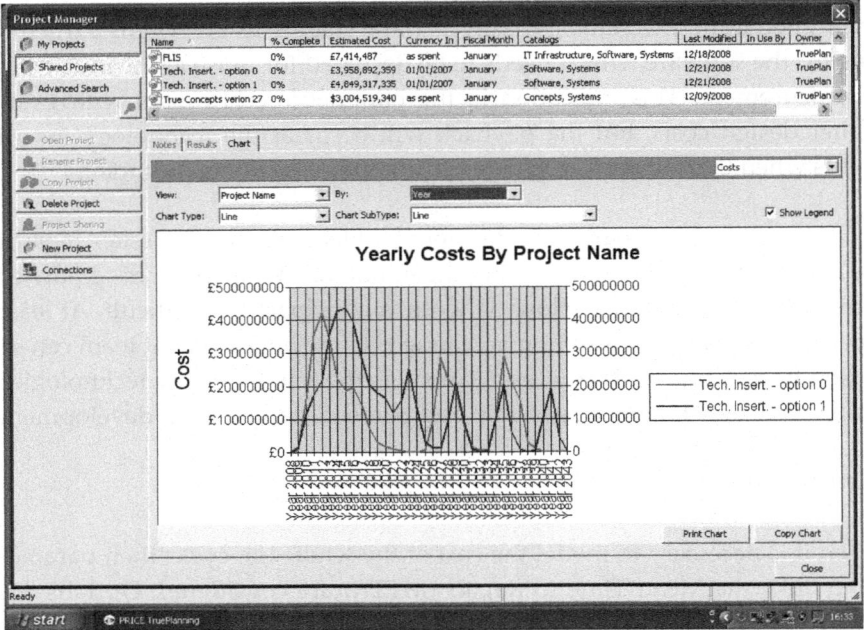

Figure 15.8 Comparison of Options 0 and 1 cost profiles

When the cost estimates are compared, it is possible to see the cost difference on an equitable basis. Both options have been calculated using the same cost model and are therefore directly comparable. For this UAV system the non-recurring costs are dominant compared to the rest of the life-cycle costs.

Even when compared on a NPV basis, in this case, there would seem to be no economic argument for selecting the technology-insertion-receptive solution. However, all the technology insertion issues have been considered together, hence further investigation may be necessary to determine if any one technology insertion issue or combination of technology insertion issues could be beneficial.

It is also worth considering the benefit as well as cost that will result from the technology insertion. In this case study the financial axis has been determined by cost modelling, however, there is the prospect for an increased average capability to be sustained through the life of the program. This non-financial measure could be dominant and the reason for considering a program which adopts technology insertion.

Summary

This parametric solution demonstrates the flexibility of parametric cost models to assist in solving engineering problems. It is not possible to construct two identical systems, one with and one without the benefits of technology insertion to monitor their cost benefits through life. However, using a parametric model it is not onerous to model the two programs with their engineering solutions and compare their whole life costs on an equitable basis. Thus, a direct comparison of these two proposed programs can be judged.

In this chapter you have learnt that acquisition and procurement programs can consider the issues around upgrade and upkeep. You have also learnt that a parametric cost model can enable you to simulate their application in the real world and the affect that it might have on the funding profile, and overall cost of a program.

16
How to ... Develop Cost-effective Alternatives

The first statement regarding Program Affordability Management in Chapter 1 of this book is: 'no program will ever be conceived without a credible analysis of alternatives.' It is important to consider the technical and business options in any proposal, bid or response to a Request for Proposal (RFP). However, it is all too common for organizations to present their traditional offering without bearing in mind what alternatives could be considered, this is especially important when budgets are tight and funding difficult.

A good Cost Engineer should seek the 'exam question'. It is too easy just to start estimating and jump into action, only to find later that the estimate was not required, or the cost generated did not satisfy the problem. It is not the job of the Cost Engineer to interpret the RFP, but with common sense and cost estimating specific questions it is possible to direct the proposal or bid process to a more successful conclusion to the satisfaction of all.

Seek the Concept of Analysis (COA) or a similar document. This should state what options are going to be considered in response to a study or RFP. It should state clearly and categorically what the exam question is, for example, 'the customer is seeking XX vehicles of YY capability with a budget of ZZ'. It should be possible to eliminate some option in the document itself without the need for estimates, for example, some vehicle solution might simply be too heavy or too large for the capability this customer requires – so there is no need to waste time estimating these.

It is worth determining the type of analysis that is required. In the COA it should state what the customer is going to receive. It is too common at the end of the process for fingers to be pointed if the bid is unsuccessful. Ideally, senior management will sign up to the COA at the start of the process, then when the proposal and estimate is presented, there will be no last minute surprises, and with luck no need to burn the midnight oil.

Ask the questions: Is this a committing proposal or a Rough Order of Magnitude to enable the customer to set budgets? Does this need more than one estimating methodology to be employed or will one be sufficient? Is this estimate the outcome of a study or the committing bid for a derivative to an existing product? It is important to commit resource relative to the importance of the estimate.

The COA should guide the process into an economic analysis or a financial analysis. The financial analysis will determine the affordability of a project and therefore it is important that the accuracy of the estimate is good. Other financial analyses may require a commercial style accounting system applying interest on capital and depreciation to fixed and sunk costs.

This compares to an economic analysis in which options can be compared. In this case the relative cost of the option to each other is important and the total cost might not be relevant. The modelling will generally only consider marginal costs for an Investment Appraisal (IA) and will exclude any fixed, common and sunk costs. Fixed costs will not be affected by the number of systems procured, for example, the cost of runways, docks or launch pads. Common costs will not be affected by the technical design differences between the options, for example, the cost of hardware in a software study. Sunk costs will not be affected by the option chosen as they are spent or already committed, for example, previous expenditure on concept phases or studies. A great deal of time and effect can be saved just by getting the ground rules straight.

Figure 16.1 shows the top level process, starting from the RFP, Statement of Work or technical study output, which then leads to a business analysis and technical analysis. These activities will result in a COA and technical proposal or technical study respectively. The real estimating element begins with a review of these documents to propose a Product Breakdown Structure (PBS) and risk register.

The next sensible task is to issue a Master Data and Assumptions List (MDAL) or Cost Analysis Requirement Description (CARD) to frame the estimate to be created. Once the data has been gathered, this can be interpreted by the Cost Engineer into the cost model and costs estimated. With the addition of uncertainty following a risk identification activity, it is possible to put a

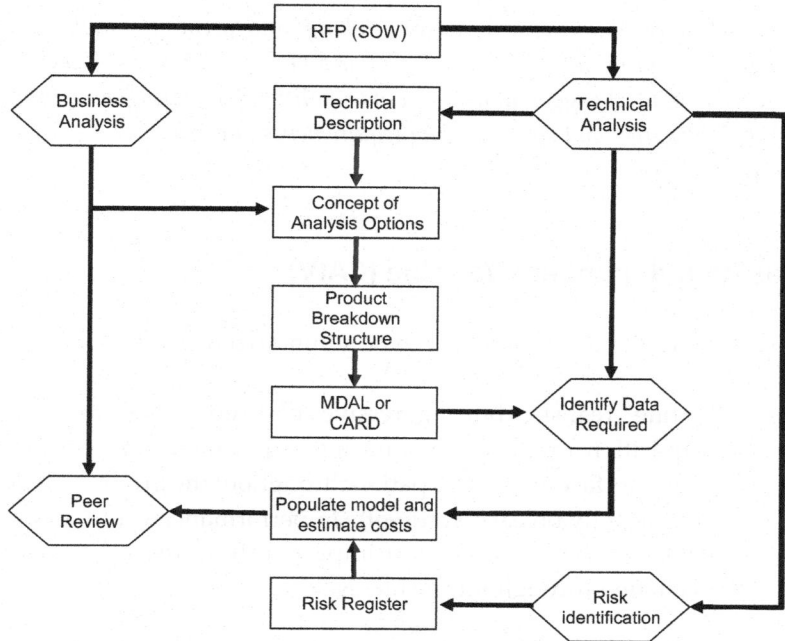

Figure 16.1   The top level process

tolerance on the estimate and a peer review can be held prior to final delivery of the estimate.

One of the most successful campaigns to promote the consideration of alternatives has been the Cost as An Independent Variable (CAIV – pronounced 'cave') initiative of the US Department of Defense (DoD). When Cost Engineers have looked to influence the design process, this has taken the traditional form of Design to Cost, however, this tends to be defined as design to production cost rather than Whole Life Cost. The CAIV initiative focuses on the Whole Life Cost (WLC) as it is possible that increased expenditure during the production phase might actually reap rewards in a lower overall WLC.

Cost as An Independent Variable (CAIV) is defined as:

> *Methodology used to acquire and operate affordable DoD systems by setting aggressive, achievable Life Cycle Cost (LCC) objectives and managing achievement of these objectives by trading off performance and schedule, as necessary. Cost objectives balance mission needs*

with projected out-year resources, taking into account anticipated process improvements in both DoD and industry. CAIV has brought attention to the government's responsibilities for setting/adjusting LCC objectives and for evaluating requirements in terms of overall cost consequences.[1]

Cost as An Independent Variable (CAIV)

The objectives of the CAIV process can be summarized as follows:

- Setting realistic but aggressive cost objectives early in each acquisition program. This means acquiring strong customer and user interface and participation throughout the program. It depends significantly on early trade-offs in performance versus costs. These trade-offs have the greatest impact early in the program and are evaluated throughout the life-cycle.

- Managing risks to achieve cost, schedule and performance objectives. This means setting early, aggressive, yet realistic, goals and adjusting them when necessary. It also means using risk analysis models to give decision-makers the insight into changes to cost, schedule and performance goals, how these changes affect life-cycle costs, and how decision-makers can quantify and mitigate risk.

- Devising appropriate metrics for tracking progress in setting and achieving cost objectives. This means setting management targets and goals that are realistic. It also means the ability to acquire history and then use that history to evaluate future activity.

- Motivating government and industry managers to achieve program objectives. This means generating incentives for the contractor and providing the customer with optimized solutions. A CAIV compliant organization has the ability to pursue and realize greater incentives.

- Putting in place for fielded systems additional incentives to reduce Operating and Support costs.

---

[1] http://www.dau.mil/pubs/glossary/12th_Glossary_2005.pdf, *Glossary of Defense Acquisition Acronyms and Terms*, 12th edition, July 2005.

So what is CAIV? It has already been established that the focus is on the whole life-cycle, but what more? To start with, the customer is expected to participate in the whole life-cycle, not just write a specification at the start and await delivery at the end. This enables challenging-of-requirements dialogue between customer and contractor. It becomes a team philosophy which ensures that cost becomes equal in importance to design and performance parameters when making program decisions. The process includes the establishment of systems costs based upon mission affordability, which leads to cost targets for individual system elements being brought down from these affordability goals.

During the design phase the user's requirements and the costs are considered together and with equal importance. The most successful approach is to integrate the requirements analysis, design and costing tools to aid this process, not just for the initial design, but through the life of the project to ensure that corrective actions can be taken if necessary. Risk analysis is an integral part of the process to ensure that the program is not knocked off course and that results are tracked and measured in the form of metrics.

Put simply, CAIV is a process which seeks to reduce costs, while Life Cycle Costing provides the domain, a set of costs to be reduced.

Trade-Off Studies

So how is this possible? What can be achieved? In simple terms, the solution is the integration of performance data in order to establish performance to technology relationships. In the case of parametric cost models, this is easy as the technology index is the Manufacturing Complexity, so a performance to Manufacturing Complexity relationship needs to be established.

Figure 16.2 shows the trade studies process in simple terms; this is an iterative process aiming to achieve a target cost. The process begins with consideration of the mission to be achieved, design anticipated and date that the system is required. These details can be interpreted into a parametric cost model, with the performance being a driver of the Manufacturing Complexity.

Once the inputs have been populated, the cost engine of the parametric model is stimulated and the cost estimate created. As described earlier, this will be a Whole Life Cost estimate including the Operating and Support costs.

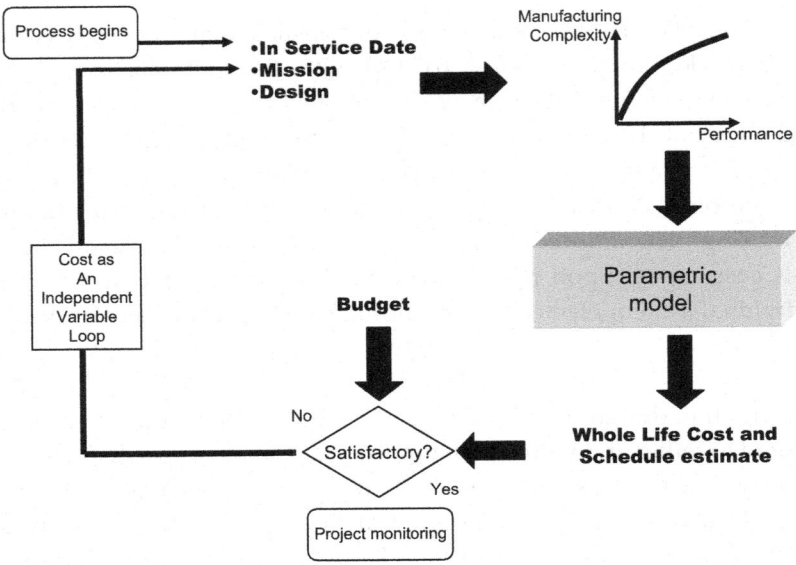

Figure 16.2 Trade studies process

This cost estimate is compared to the budget available to the project. If this technical design has resulted in a solution and estimate that is compatible with the budget, all is well. If, however, the cost or schedule estimate exceeds the budget or In-Service Date, then a CAIV loop is invoked and the design must be adjusted.

This process depends on the establishment of the performance to Manufacturing Complexity relationships prior to embarking on the exercise. It is too late in the pressure of a bid or proposal to start the process of determining these relationships; forethought and planning of a well organized cost engineering department will realize the benefit of this exercise before the need arises.

Figure 16.3 shows a relationship for gas turbines, where thrust in pounds has been related to the Manufacturing Complexity of the gas turbine. Engineering logic indicates that the more thrust required, the more complex the technology will be to deliver it. The data in this graph backs up this hypothesis.

Cost as An Independent Variable (CAIV) entails setting aggressive, realistic cost objectives for acquiring systems, and managing program risks

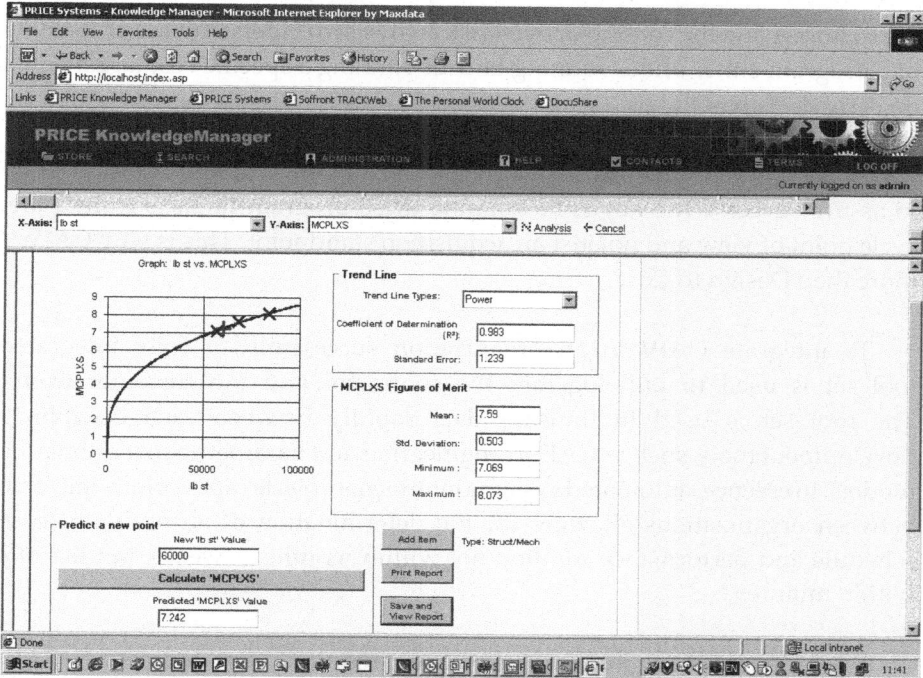

Figure 16.3 Performance driven gas turbine complexity calculator

to achieve those objectives. Cost objectives must balance mission needs with projected resources, taking into account existing technologies as well as new technologies.

Implementation

Achieving a CAIV process means having the right analytical support and tools for developing an optimized CAIV program and tracking it through its life-cycle. This requires a thorough understanding of the risks involved from choices in the early concept phase and their effect on total life-cycle.

CAIV analytical tools must not only provide the capability of optimizing design, cost and schedule, but also provide decision-makers the capability of assessing and mitigating risk throughout the program's life-cycle. For example, program managers need to understand how trading among cost, performance and schedule may impact a program's ability to stay within cost targets. Both the customer and the contractor need risk analysis to understand the probability of a program either under-running or over-running a specific cost target based on early decisions.

This leads to results-oriented corrective action early enough in the program to assure success. The key is to examine these program impacts from a life-cycle point of view and not just an acquisition standpoint. This is why CAIV is more than Design to Cost.

To integrate CAIV in an organization successfully, a fully integrated tool set is used to pull together the contractor and the customer teams. The tool set is used to transmit data rapidly between each discipline's development tools, such as CAD in engineering and parametric cost estimating models. In essence, automated systems interfaces provide rapid communication between organizations allowing for the determination of an optimum cost, schedule and performance solution not within months or weeks, but literally within minutes.

While automated interfaces are not required for successful CAIV implementation, they allow the engineering and business communities to iterate hundreds or thousands of solutions in a short time period and instantly communicate the results at all levels of the Integrated Product Development teams. The end result of this process is the capability to make informed design decisions today that will ensure system affordability tomorrow.

CAIV requires that we continually balance the program to optimize cost/performance and schedule. The integration and balancing of the essential issues are accomplished by:

- A CAIV committed organization where CAIV principles are incorporated at the highest levels and cover the essential functions within the engineering and business departments. Commitment from top management is essential for CAIV success. This includes the seamless involvement of both the engineering and estimating organizations.

- It also means having a team made up of the design and performance groups necessary to determine technical feasibility within the context of CAIV. Part of assessing technical feasibility means the ability to trade-off cost, performance and schedule and communicate the optimized results to both the contractor and customer. It also means having the tools to analyse the interactions between these complex issues.

- Risk analysis is required to understand the impacts of trading among cost, schedule, technology and performance. The results of risk analysis provide decision-makers with the information required to mitigate risk and examine the probabilities associated with achieving program results.

- CAIV tools include parameter-based cost models for hardware and software. They also include enterprise models that allow complete and seamless integration between engineering CAD tools and parametric cost models so that design and performance parameters can instantly be assessed, risks mitigated, and information provided in an appropriate time-frame to the decision-makers.

Let us now look further at CAIV implementation and a process for enabling an organization with a CAIV business strategy.

Trade Study Opportunities

At this point it will be necessary for some organizations to undergo a complete change of philosophy. The trade study process for CAIV requires that cost-performance trade studies are used as a design tool rather than confirmation of selected concepts. Hence the need for parametric models to provide rapid life-cycle cost analysis.

As such, cost thinking must start at the engineering level. Engineers must understand the importance of treating cost as equal to design and performance and, in essence, treat cost as another design parameter. In treating cost as another design parameter and using a parametric cost model, engineers need to understand how a parametric model is influenced by key parameters such as:

- Specification level – where does the product operate?

- Complexity of the engineering tasks – what is the skill mix and what type of design, that is, new/modified/repeated or off the shelf?

- Technology – what is the current and forecasted technology? What is the optimized technology mix that balances cost and performance?

- How quantity affects programs – this influences not only unit production cost but also Operation and Support (O&S).

- Schedule – how does schedule influence production rate and therefore cost? How can schedule influence tooling costs?

- How the engineering parameters also influence interaction with: maturity, precision, technology, schedule, manufacturing and material (number of parts).

For example, in the production and assembly process, we need models that can determine the impact of technology and cost reductions for reducing part count, and improving manufacturing techniques and Mean Time Between Failures (MTBF). Figure 16.4 provides a simple representation of the trade studies, but all these impacts need to be examined from the concept phase through to the Operation and Support phase.

A CAIV enterprise strategy consists of an organization concurrently using both parametric models and system engineering tools. These tools may be automatically linked through an Integrated Design to Cost (IDTC) approach. A fully realized IDTC tool-set allows every functional department within an organization to integrate seamlessly into the CAIV process, as shown in Figure 16.5. Information is shared and analysed in real time, thus providing

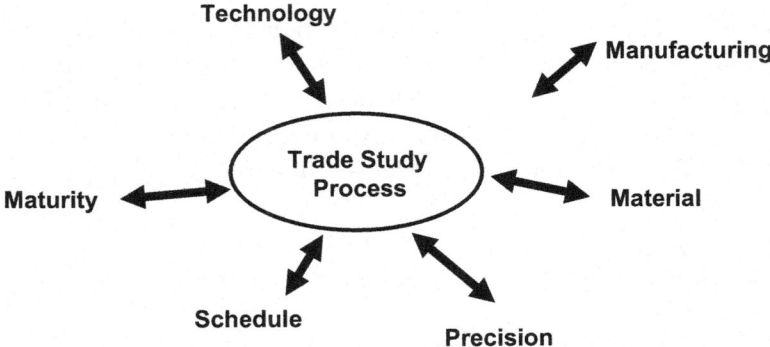

Figure 16.4 Trade study considerations

Figure 16.5   Enterprise integration

decision-makers with crucial data in minutes rather than hours, days or weeks. This process starts during concept design and continues through Operation and Support. The impacts of value engineering and the trade study process are well communicated through the organization and the customer.

Summary

Cost as An Independent Variable (CAIV) is not just an American initiative, but good sound business practice facilitating:

- optimized consideration of cost, schedule and performance;
- the evaluation of CAIV competencies and areas for improvement;
- the promotion of customer and contractor communication;
- truly meeting the end user's needs;
- the realization of greater incentive award fees.

The CAIV trade study process plays an important role in the overall process:

- trade studies occur throughout the life-cycle of a program;

- trade studies start as early as modernization planning and continue through fielding and service life extension;

- trade studies involve a wide range of tools, including:

    - parametric cost models;
    - cost management;
    - optimization tools;
    - risk models;
    - campaign models;
    - decision support tools.

17

How to ... Tackle the System of Systems Challenge

In the 1980s a number of major contractors made the strategic decision to become the Design Authority on Aerospace and Defence Systems. Traditionally, suppliers had provided a service to fulfil the whole spectrum of tasks from product or program conception, to design, to providing through manufacture and tests. The new strategy would mean that the large players in the industry would outsource the manufacturing of equipment and components to small and medium-sized organizations who would provide these items at more competitive prices thanks to economies of scale and demand across multiple customers. The prime contractor then assumed the risk of the equipment that they acquired, working as part of the system. They were responsible for the design integration as well as the physical system integration and testing.

Figure 17.1 shows how a parametric model structures typical system architecture. The process opens in the top-left of the illustration with the creation of a project team – starting with the appointment of a project manager, who initiates and plans the project, through to the appointment of the technical team who will take the requirements and analyse them to determine the top-level system design.

Once this has been achieved the components of the system are known and all that is required is to establish the procurements philosophy. This defines the acquisition strategy: what is to be made in-house (due to its unique characteristics), which elements can be bought off-the-shelf and what needs to be build-to-print (see the lower portion of the 'V' diagram).

Having successfully acquired these elements the systems house can integrate them and test their solution against the customer's need. Ultimately, the project team needs to validate and verify that the systems delivered by the

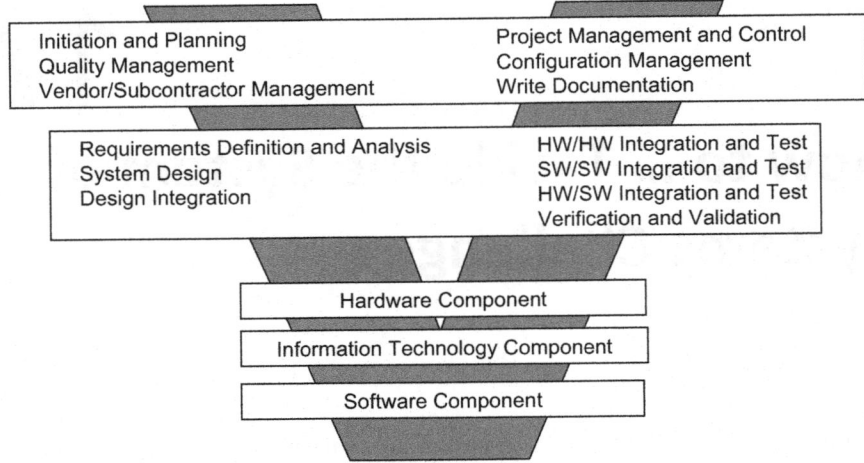

Figure 17.1 System of equipments

suppliers and any in-house development satisfy the customer's requirements prior to winding up the project.

The end-user customer thus retained the risks associated with ensuring that different projects worked together. Moreover, Government procurement organizations had the responsibility of ensuring that the interfaces (both software and hardware) between projects were adequately defined.

The structure has evolved since the turn of the century: Government organizations are looking to mitigate the risk of projects not working together by introducing the concept of System of Systems projects. They have sought systems integrators who are willing to accept the risks and the responsibility for a project and the resulting systems working together in harmony (see Figure 17.2).

Many Government defence departments have migrated from a platform-based acquisition strategy to one focused on delivering capabilities. Instead of delivering a fighter aircraft or an unmanned air vehicle, contractors are now being asked to deliver the right collection of hardware and software to meet specific wartime challenges. This means that much of the burden associated with conceptualizing, designing the architecture, integrating, implementing, and deploying complex capabilities into the field has shifted from desks in Government to desks at Lockheed Martin, Thales, BAE Systems, EADS, Boeing, and other large aerospace and defence contractors.

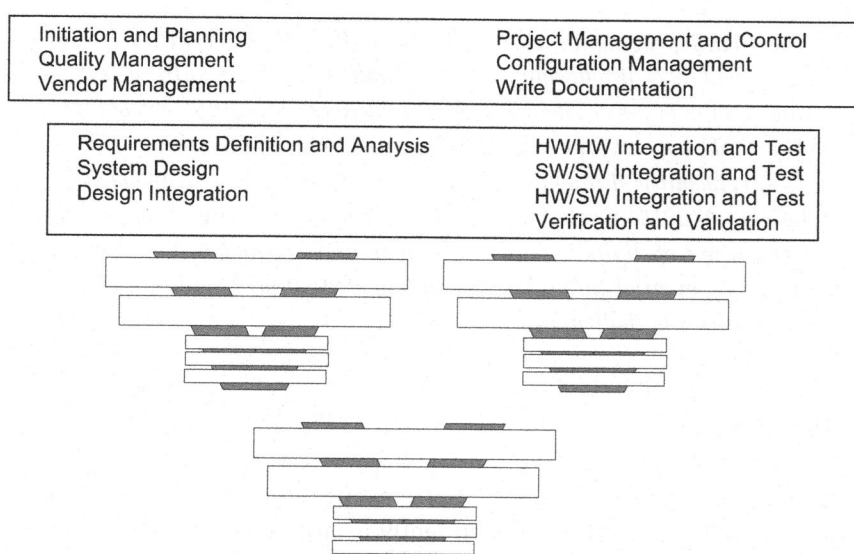

Figure 17.2   System of Systems project structure

In 'The Army's Future Combat Systems' [FCS] Features, Risks and Alternatives', the Government Accounting Office describes the challenge[1]:

> ... 14 major weapons systems or platforms have to be designed and integrated simultaneously and within strict size and weight limitations in less time than is typically taken to develop, demonstrate, and field a single system. At least 53 technologies that are considered critical to achieving critical performance capabilities will need to be matured and integrated into the system of systems. And the development, demonstration, and production of as many as 157 complementary systems will need to be synchronized with FCS content and schedule.

The future in the UK is exemplified in the House of Commons Defence Committee's report 'The Army's Requirement for Armoured Vehicles: The FRES Programme',[2] which states:

---

1 Paul L. Francis, 'The Army's Future Combat Systems' Features, Risks and Alternatives'. Testimony before the Subcommittee on Tactical Air and Land Forces. Committee on Armed Services. House of Representatives, GAO-04-635T 1 Apr. 2004 <www.gao.gov/new.items/d04635t.pdf>.
2 House of Commons Defence Committee, 'The Army's Requirement for Armoured Vehicles: The FRES Programme', Seventh Report of Session 2006–07, Published 21 Feb. 2007. <http://www.publications.parliament.uk/pa/cm200607/cmselect/cmdfence/159/159.pdf>.

The FRES programme is expected to deliver 3,000 vehicles in 16 battlefield roles. It will comprise three families of vehicles: utility, heavy and reconnaissance. The MoD plans to deliver the utility vehicle first. The MoD has set four key requirements of the FRES utility vehicle: 1. survivability through the integration of armour; 2. deployability by the A400M aircraft; 3. network-enabled capability through the integration of digital communication technology and 4. through-life upgrade potential throughout its anticipated 30-year service life. The requirement is challenging.

What is a SOS?

In the defence community a SOS (System of Systems) is a configuration of component systems that are independently useful but synergistically superior when acting in concert. In other words, it represents a collection of systems whose capabilities, when acting together, are greater than the sum of the capabilities of each system acting alone.

Today, there are many platforms deployed throughout the battlefield which only have limited means of communication. This becomes increasingly problematic when multiple services are deployed on a single mission because it means that there is no consistent means for the Army to communicate with the Navy, or the Navy to communicate with the Air Force. Inconsistent and unpredictable means of communication across the battlefield often result in an unacceptable time from detection of a threat to engagement. This is further exacerbated when acting with other forces, such as NATO peacekeeping troops, whose systems are equally incompatible; strikes from friendly fire are more likely to occur in these operations.

One example of a SOS that the Army is currently envisioning is the Warfighter Information Network-Tactical (WIN-T), which is a communication system designed for reliable, secure, and seamless video, data, imagery and voice services to enable decisive, real-time combat actions. This SOS promises full, two-way communication between platforms and across services, making it possible for information to be shared and processed in time to make a real difference in the outcome.

For an example from the UK, consider the Defence Industry Strategy (DIS)[3] which proposed:

> *The Military Afloat Reach and Sustainability (MARS) program is a significant planned investment in a new integrated approach to Afloat Support, combined with investment in life extensions for retained platforms. The MARS system-of-systems may include Fleet Tankers, Joint Sea Based Logistics and Fleet Solid Support vessels.*

How Different are SOS Projects?

How different is a project intended to deliver a SOS capability from a project that delivers an individual platform such as an aircraft or a submarine? Each case presents a set of customer requirements that need to be elicited, understood and maintained. Based on these requirements, a solution is crafted, implemented, integrated, tested, verified, deployed and maintained. At this level, the two projects are similar in many ways. Dig a little deeper and differences begin to emerge. The differences fall into several categories: acquisition strategy, software, hardware and overall complexity.

ACQUISITION STRATEGY

The SOS acquisition strategy is capability-based rather than platform-based. For example, the customer presents a contractor with a set of capabilities to satisfy particular battlefield requirements. The contractor then needs to determine the right mix of platforms, the sources of those platforms, where existing technology is adequate, and where invention is required. Once those questions are answered, the contractor must decide how best to integrate all the pieces to satisfy the initial requirements. This capability-based strategy leads to a project with many diverse stakeholders. Besides the contractor selected as the Lead System Integrator (LSI), other stakeholders that may be involved include representatives from multiple services, Defence Advanced Research Projects Agency and prime contractor(s) responsible for supplying component systems as well as their subcontractors. Each of these stakeholders brings to the table different motivations, priorities, values and business practices – each brings new people management issues to the project, too.

---

3   Defence Industrial Strategy (DIS), Defence White Paper, December 2005. <http://www.mod.uk/NR/rdonlyres/F530ED6C-F80C-4F24-8438-0B587CC4BF4D/0/def_industrial_strategy_wp_cm6697.pdf>.

SOFTWARE

Software is an important part of most projects delivered to customers. In addition to satisfying the requirements necessary to function independently, each of the component systems needs to support the interoperability required to function as a part of the entire SOS solution. Much of this interoperability will be supplied through the software resident in the component systems. This requirement for interoperability dictates that well-specified and applied communication protocols are a key success factor when deploying a SOS. Standards are crucial, especially for the software interfaces. Additionally, because of the need to deliver large amounts of capability in shorter and shorter timeframes, the importance of Commercial Off the Shelf (COTS) software in SOS projects continues to grow.

With platform-based acquisitions, the customer generally has a fairly complete understanding of the requirements early on in the project, with a limited amount of requirements growth once the project commences. Because of the large-scale and long-term nature of capability-based acquisitions, the requirements tend to emerge over time with changes in governments, policies and world situations. Because requirements are emergent, planning and execution of both hardware and software contributions to the SOS project are impacted.

HARDWARE

System of Systems (SOS) projects are also affected by the fact that the hardware components being used are of varying ages and technologies. In some cases, an existing hardware platform is being modified or upgraded to meet increased needs of operating in a SOS environment, while in other instances brand new equipment with state-of-the-art technologies is being developed. SOS project teams need to deal with components that span the spectrum from the high-tech, but relatively untested to the low-tech, tried-and-true technologies and equipment.

COMPLEXITY

Basically, a project to deliver a SOS capability is similar in nature to a project intended to deliver a specific platform, except that overall project complexity may be increased substantially. These complexities grow from capability-based acquisition strategies, increased number of stakeholders, increased overall cost (and the corresponding increased political pressure), emergent requirements, interoperability, and equipment in all stages from infancy to near retirement.

New and Expanded Roles and Activities

Understanding the manifestation of these increased complexities in a project is the first step to determining how the planning and control of a SOS project differs from that of a project that delivers one of the component systems. One of the biggest and most obvious differences in the project team is the existence of a Lead System Integrator (LSI). The LSI is the contractor tasked with the delivery of the SOS that will deliver the capabilities the customer is looking for. The LSI can be thought of as the super-prime or the prime of prime contractors. They are responsible for managing all the other primes and contractors and ultimately for fielding the required capabilities. The main areas of focus for the LSI include:

- requirements analysis for the SOS;
- design of SOS architecture;
- evaluation, selection and acquisition of component systems;
- integration and testing of the SOS;
- modelling and simulation;
- risk analysis, avoidance and mitigation;
- overall program management for the SOS.

One of the primary jobs of the LSI is completing the system engineering tasks at the SOS level (see Figure 17.3).

Focus on System Engineering

System engineering as a discipline first emerged during the Second World War as technology improvements collided with the need for more complex systems on the battlefield. As systems grew in complexity, it became apparent that it

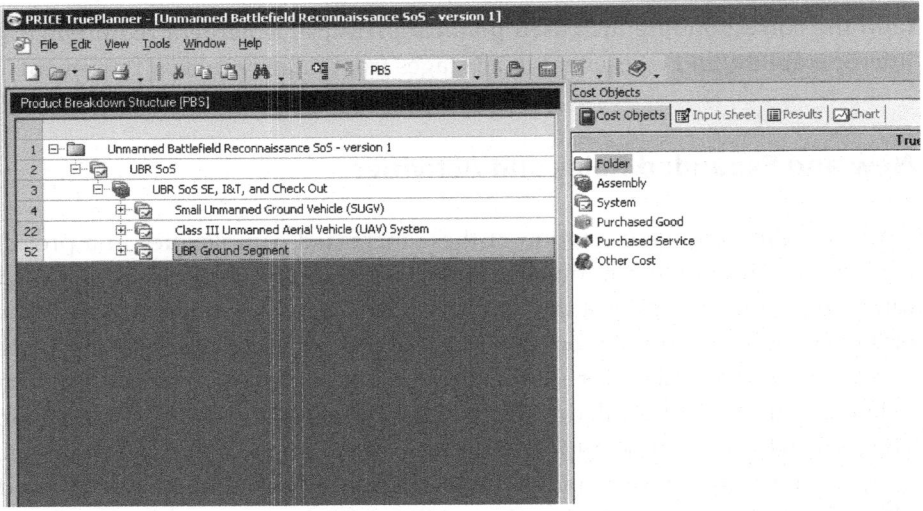

Figure 17.3   Lead System Integrator (LSI) and primes

was necessary for there to be an engineering presence well versed in many engineering and science disciplines to lend an understanding of the entire problem a system needed to solve. To quote Admiral Grace Hopper, 'Life was simple before World War II. After that, we had systems.'[4]

With a top-level view, Systems Engineers were able to grasp how best to optimize emerging technologies to address the specific complexities of a problem. Where an Electrical Engineer would create a solution focused on the latest electronic devices and a Software Engineer would develop the best software solution, the Systems Engineer knows enough about both disciplines to craft a solution that obtains the best overall value from technology. Additionally, the Systems Engineer has the proper understanding of the entire system to perform validation and verification upon completion, ensuring that all component pieces work together as required.

Today, a new level of complexity has been added with the emerging need for SOS, and once again the diverse expertise of the Systems Engineers is required. Systems Engineers need to comprehend the big picture problems whose solution is to be provided by the SOS. They need to break these requirements down into the hardware platforms and software pieces that best deliver the desired capability, and they need to have proper insight into the

---

4   'The Wit and Wisdom of Grace Hopper', *OCLC Newsletter* (167) (March/April 1987).

development, production and deployment of the component systems to ensure not only that they will meet their independent requirements, but also that they will be designed and implemented to properly satisfy the interoperability and interface requirements of the SOS. It is the task of the Systems Engineers to verify and validate that the component systems, when acting in concert with other component systems, do indeed deliver the necessary capabilities.

Cost Considerations of SOS Projects

A SOS is a collection of existing, upgraded and new systems that are required to work together to accomplish specific objectives. Clearly the costs of developing and acquiring component systems is one important cost consideration, but since estimating system costs is a fairly mature discipline, this chapter focuses on the additional costs associated with the delivery of capabilities made possible when a configuration of such systems works as a system.

Mastering the cost questions in a SOS project first requires establishing a link between the increased complexities and the participation of Systems Engineers in the project. A traditional parametric estimating methodology for hardware or software systems relies on a quantification of the size and complexity of the system being developed. Size is driven by weight for hardware and Source Lines of Code or Function Points for software. Project, process and organizational factors drive complexity. Assigning a size and complexity to a SOS is a bit trickier. Traditional size measures alone are not adequate for estimating the size of system engineering tasks and with many participating organizations, the process and organizational factors can vary substantially within the project team.

Tricky or not, being able to properly size the SOS part of a project is crucial to successfully determining what it will cost and how long it will take to deliver. It is also a crucial step in being able to make trade-offs in order to deliver a solution that not only meets requirements, but also satisfies affordability constraints. As with all estimating, the challenge in sizing a SOS is being able to translate what is known early on in the project into information that represents useful project characteristics as it evolves.

Toward this end, research indicates that the number of unique interface protocols and the number of different component systems are the two best factors for determining the size of the SOS effort. In a SOS project, it is the LSI's job to define and design the infrastructure that will facilitate communication among the many component systems. The number of unique interface protocols is clearly a good start for determining problem space size. Augmentation of this number with the number of component systems that will be designed for or adapted to operate within this infrastructure provides an even better proxy for the size of the solution. This conclusion is consistent with the research conducted at the University of Southern California's Centre for Software Engineering on the Constructive System of Systems Integration Model.[5]

The number of unique interface protocols drives the size of the integration and test effort. The effort for integration and test within a typical system ranges between 5 and 40 per cent of the entire development effort of the system as the number of interfaces goes from few to many; this effect would be exaggerated in a SOS as complexity of the overall integration problem is greater. As the number of component systems increases, integration efforts increase in a non-linear fashion as a result of the diseconomy of scale brought on by project complexity. Additionally, the number of components will influence management and oversight costs in the form of added people and communication issues.

Size, of course, is only part of the puzzle. Multiple SOS within the same size range will only fall into the same cost range as a coincidence. For the sake of this discussion, consider the simplistic cost model that applies an exponent and a coefficient to a project size. In this context, the size is as described and the exponent and coefficient are determined by factors that determine project complexity. As such, it is necessary to assign relative complexity values to the various configurations. There are many factors that have a potential impact on complexity, some that are obvious early on in the project and others that will emerge throughout the project life-cycle. The ones that are available or predictable early on in the project, and that appear to have the most significant impact on the amount of effort required for the SOS tasks, include those below.

NUMBER OF OPERATIONAL SCENARIOS

An operational scenario refers to a particular capability instance for some set of the component systems of the SOS. For example, the Coast Guard's Integrated Deepwater System needs to include capability that can react to a

---

5   Jo Ann Lane, 'Factors Influencing System-of-Systems Architecting and Integration Costs', University of Southern California's Center for Software Engineering, <www.stevenstech.edu/cser/authors/46.pdf>.

terrorist threat, a person lost at sea or a drug-smuggling operation. The number of operational scenarios impacts the coefficient in the cost equation discussed earlier as additional scenarios result in more time for requirements, design and modelling and simulation. Depending on the similarities of the scenarios, the impacts to these activities' costs should represent increases between 10 and 50 per cent.

REQUIRED LEVEL FOR ACCEPTANCE OF KEY PERFORMANCE PARAMETERS

Key performance parameters associated with a SOS include things like detection effectiveness, survivability and lethality. This factor could have substantial impact on both the coefficient and the exponent in the simple cost model mentioned earlier. System engineering activities associated with the SOS could double or triple, even more as the detection effectiveness expectations move from available technology to state-of-the-art. Use of immature technology on the Joint Tactical Radio System Program was cited as one of the main reasons for a $458 million development cost increase.

NUMBER OF SUPPLIERS AND STAKEHOLDERS

The number of players involved in a SOS project can increase the complexity and cost significantly. On a typical system project, people and communication issues can increase the cost of project management and oversight activities by as much as 60 per cent. This effect can increase dramatically as the relatively well-known confines of the typical system are replaced with the much more expansive and undefined constraints on a SOS project.

INTEGRATION COMPLEXITY

Integration complexity is a quantification of the amount of integration each component is expected to require with the rest of the SOS. A SOS that requires highly complex integrations within and among each of its component systems could potentially see the integration and test activity costs increase by an order of magnitude from a SOS where all of the integrations are simple, well-defined tasks.

STABILITY AND READINESS OF COMPONENTS

As mentioned earlier, the technical immaturity of components can substantially impact system engineering tasks. Additionally, immature components can impact the overall schedule and cost of the SOS, since integration and test activities for various capabilities will be delayed until all required components are available. The Warfighter Information Network-Tactical (WIN-T) program was originally planned to deliver technologies not expected to mature until after production started. Such a strategy is guaranteed to lead to costly schedule delays.

AMOUNT OF COTS CAPABILITY

COTS components generally require modification, integration and testing, as well as compromise on SOS requirements. When looking at the overall cost for a SOS, off-the-shelf components should decrease the cost compared to newly developed components. From the perspective of the LSI, however, they represent an increase of system engineering effort associated with requirements, design and integration, and testing. This cost increase can be quite modest if the components and vendors are chosen wisely, but it could double the costs of these activities if poor choices are made.

Affordable SOS

Armed with this Cost Driver knowledge when crafting a solution to deliver a SOS capability, there are things the LSI can do to ensure that it not only meets all performance requirements, but does so within affordability constraints. All possible solutions should be focused on the specified constraints for stated Key Performance Parameters (KPPs). No solution should be presented that does not satisfy these constraints. Component systems that drive performance substantially above specified performance in these areas should be carefully scrutinized as well. All possible solutions should first be validated to ensure that they successfully address all KPPs and support all operational scenarios.

Care should be taken to use as many existing component systems as possible rather than developing new ones. When new component systems must be developed to deliver some currently non-existing capability or degree of performance, it is important to extract the most from the technology investment. Attempts should be made to incorporate as much capability as practical into the new development to reduce the number of different component systems. Increases in complexity associated with technology readiness and component stability may be offset by size decreases if the number of required component systems can be reduced. At the same time, care should be taken to ensure that expectations for technology do not exceed practical limits on innovation imposed by schedule constraints on the program.

Well thought out architecture with simple communication protocols that meet many different needs will reduce the size of the SOS solution space. Although there is an up-front investment in getting the architectures right and standardizing communication protocols, the payoff is significant during delivery of the initial operating concept and throughout the life of the SOS. Emerging requirements will result in the addition of new component systems that must communicate with existing components.

The use of COTS hardware and software is a practical and necessary approach to accomplish the delivery of SOS capabilities within required timeframes. When possible, the same vendor should be considered for multiple components, parts or software products. This reduces the number of vendors involved in the project, eases the effort to integrate between components, and could possibly result in favourable purchasing agreements based on bulk. Integration complexity can also be reduced through simple standards that are strictly enforced, effective risk management techniques focused on early identification and mitigation, and ongoing integration efforts.

Summary

Today, SOS solutions are replacing the existing post-Second World War systems as the next generation of complex solutions supplied by contractors to Government defence departments. SOS projects require contractors to deliver capabilities rather than stand-alone systems. Contractors are left to decide on and acquire component systems, determine the best configuration for these component systems to achieve the required capabilities, and develop the best plan for interoperability among the component systems.

While there are some ways in which a SOS project is similar to a project that delivers a component system, there are many ways in which the two types of project differ. Understanding these differences and how they affect the cost and effort associated with a project is crucial to proper planning and execution of a

SOS project. A crucial difference is the requirement for an increased involvement of system engineering resources throughout the life-cycle of the SOS project. Systems Engineers are involved in requirements elicitation and management, architecture decisions, test and evaluation, verification and validation, and technical oversight for the SOS project.

It is necessary to begin estimating these projects today by incorporating estimating knowledge gained through years of system development augmented with information about the additional factors that influence SOS project size and complexities.

18

How to ... Create Home-grown Parametric Models

Since 1975 parametric model vendors have been supplying cost and schedule parametric cost models. Now users have begun a second revolution by demanding a costing system capable of storing organic, home-grown Cost, Schedule and Performance Estimating Relationships.

This chapter describes how to create a parametric cost model. It is based on a methodology, rather than a model, and uses the case study of concepts estimating. The main focus is on the cost research necessary to build a cost model. This methodology is equally applicable to any cost model that an organization can conceive.

The conceptual model will be used during high-level studies considering early concepts when little information is available to the Cost Engineer. As a result, a high level of detail cannot be expected. The outputs that are anticipated are:

- acquisition costs (development and production);
- in-service costs (Operating and Support);
- disposal cost.

The Construction Capability

It is important that procedures are employed to control the cost research and separate procedures employed for the application of the mathematical models that result from that research. These requirements are mirrored in software of a third generation parametric model that is used to aid the cost analysts in creating the models, and separate user interface software that is used by Cost Engineers to apply those same models to estimating problems.

For some time, parametric model vendors have realized that providing software alone to customers is not enough. Users also require procedures and processes to obtain the full benefit of the models that they license. For this reason parametric systems are not just software packages, but a complete parametric estimating system.

A minority of cost analysts and researchers will want to build activity-based custom cost models. They will be able to create parametric, analytical – sometimes called *grass-root* or *bottom-up* – estimates or analogous cost models. The advantages of using this framework are many, but the obvious ones include the ability to consolidate commercial and public cost models.

It is also possible to consider commercial models and alter or modify the algorithms if it is deemed that an organization has better information – for example, if the organization uses a particular form of cost improvement curve not already implemented.

The models apply a cost framework that has been designed specifically for Cost Engineering; this framework did not start out satisfying other business requirements (for example, manufacture or procurements needs) and evolve into an estimating tool. As such it is easy to create and maintain models.

Naturally, once an organization has invested resources in the underpinning research of a cost model, they only achieve a return on the investment if the models are shared with colleagues in a controlled environment. Ideally, a cost model framework would provide features including:

- client/server capability;

- common cost escalation features;

- what-if analysis;

- risk analysis;

- Basis-of-Estimate (BOE) documentation.

Utilizing the model framework means that all these features are available to the organization.

Conceptual Modelling Case Study

It has become apparent that there is a need for a suite of high-level analogous models. Some of the attributes of these models would include:

- High level modelling to provide early indications (pre-concept) of the cost, schedule and design criteria of typical systems to feed into high level studies.

- Models that are compatible with each other and provide continuity when moving from one model to another. It would be very useful if this hardware cost model would work with other cost models.

- Simpler cost models, more specific to individual systems. Before commercial cost models, organizations such as Boeing, RCA and RAND developed platform-specific cost models; it would seem appropriate to revisit this style of model, to go back to basics.

- Generic models created with predetermined Cost Estimating Relationships, but access to the coefficients and constants in the equations, providing a simple route to implementing customer's models.

- A Through Life Estimating capability – utilizing one cost modelling framework from start to finish of a project life.

The last point is emphasized in Figure 18.1, it demonstrates how a single Graphical User Interface (GUI) should be available from beginning to end of the project life for estimating. Only the detail of the cost model should change, the man–machine interface should remain the same, with the resulting saving in training and time.

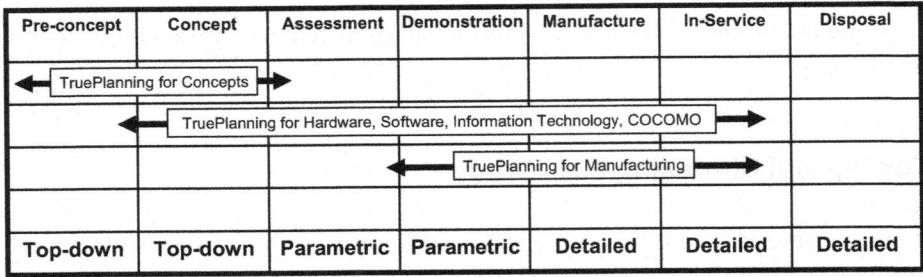

Figure 18.1 Through Life Modelling capability

CREATING A CONCEPTUAL HIGH-LEVEL MODEL

To benefit from a third generation model framework, our model needs to obey the Activity Based Costing principles – starting with a Cost Object representing the product or service. The single Cost Object can be broken down into activities that consume resources. The Cost Objects will represent a different type of platform or system, examples of those envisaged in this Catalogue are:

- unmanned air vehicles;
- large military aircraft – bomber;
- large military aircraft – transport;
- electronic counter-measures aircraft;
- reconnaissance aircraft;
- submarines – nuclear;
- submarines – conventional;
- ships – ASW and general purpose frigates;
- ships – air defence surface warships;
- missiles – air to air;

- missiles – surface to air;
- land – light armoured fighting vehicles;
- commercial – passenger;
- satellites – commercial.

COST RESEARCH

The building of a parametric cost model is quite simple (see Figure 18.2). The key to success is having a database of historical projects that have been completed and from which it is possible to draw knowledge. The important point about the database is that it must be all-encompassing; it covers all projects which could be estimated and all eventualities that could occur. The first action to take with this data is to normalize it – this is discussed in some depth below. Once normalized, consider the specific performance and design features of the projects which contributed to the resulting cost, schedule and performance of the systems.

Consider what were the inputs that influenced the resulting outputs, what are the causes and effects? The inputs are generally referred to as Cost (or schedule) Drivers. And the relationship between these Cost Drivers and the output costs are termed Cost Estimating Relationships (CERs). These are mathematical equations or algorithms that represent the influence.

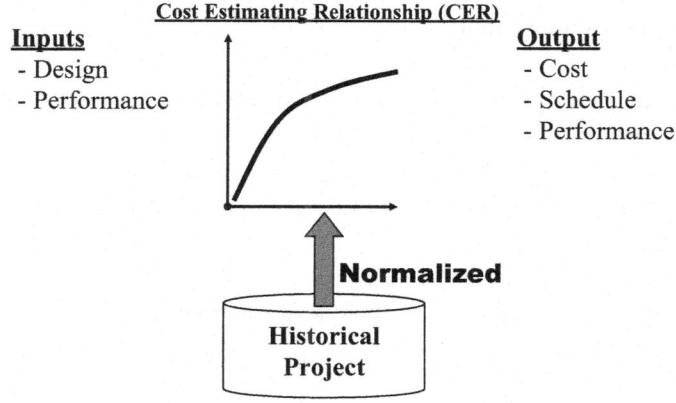

Figure 18.2   Parametric cost modelling

This is part of the cost research process; the detail is provided in Figure 18.3. The process consists of eight steps; this chapter will logically progress through these steps using the conceptual model as a case study. The process is based on the proven International Society of Parametric Analysts (ISPA) methodology for validating and verifying parametric cost models.

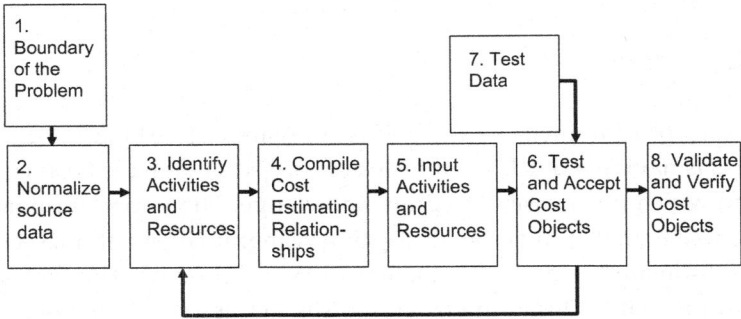

Figure 18.3   Catalogue cost research

1. Boundary of the problem

Start by articulating what is inside and what is outside the scope of the cost model that is going to be constructed. This will enable the gathering of the correct data for the model and ensuring that when the model is complete, it is possible to determine inclusions and exclusions from the estimate. It will also ensure that the data is constructed in a logical manner. For the conceptual case study, we have assumed that:

- the model will be broken down into Cost Objects – one for each system type;

- no system detail or architecture will be provided below system level;

- each type of system has a performance-driven, weight-based relationship.

2. Normalize the data

The second step in the process involves normalization of the data. This infers that the data has been collected. Eighty per cent of the time spent on this cost modelling will be consumed with data gathering. It is a time-consuming stage that should not be underestimated. Figure 18.4 provides an indication of the coverage of an appropriate database gathered for conceptual estimating. It is important that the database has historical information as well as up to date information. This database needs to be monitored and kept up to date, a database that is not maintained will quickly become obsolete and so will the cost models derived from it.

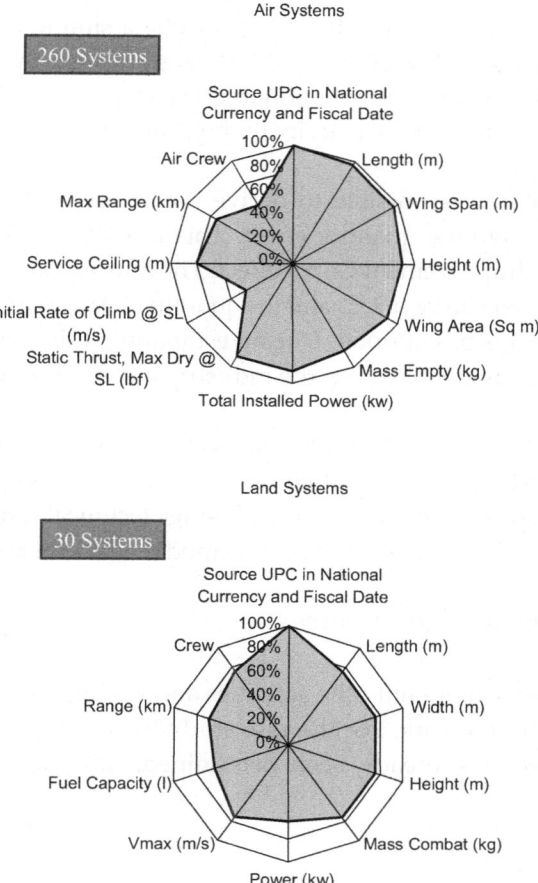

Figure 18.4   Example of a cost model database and an indication of its coverage

There are several matters to consider in normalizing the database – examples for this case study include:

- The historical project costs are first normalized for escalation, bringing the cost to a single economic value, in this case June 2006, using the Consumer Price Index (CPI) inflation table for each country where the systems were produced. It is important when storing cost data that the economic basis of the data is recorded – is the data stored as it was invoiced (referred to as 'as spent') or using constant economics? If the latter, record the base date.

- Having normalized the data for inflation, the costs are then converted into a constant currency, in this case study US dollars. The conversion mechanism could be a simple exchange rate or a Purchasing Power Parity (PPP) as published by the Organisation for Economic Co-operation and Development (OECD)[1] which also considers the productivity of the country.

- Finally, it is possible to derive the Theoretical First Piece (T_1) to eliminate the quantity effect. This is achieved by eliminating the Cost Improvement Curve effect. The information needed is the Unit Learner Curve (ULC) and the quantity that the Unit Production Cost (UPC) is based upon. During estimating, this can be implemented in the model based on the industry and environment.

This normalized T_1 cost, in 2006 US dollars ($), is our target for the cost model and could be the basis of all the other cost outputs. If this normalized historical cost can be reproduced consistently, using technical and programmatic parameters, then the basis of a parametric model has been arrived at.

3. Identify activities and resources

Next it is necessary to identify the activities and resources within the model to fit into the cost framework. For the case study the level of detail that is going to be considered has already been determined, this leads to the following activities:

- development;

---

1 www.oecd.org.

- production;
- operation;
- support;
- disposal.

As this is going to be a high-level model there are two resources to consider: labour and non-labour (material and other direct costs combined).

At this level of modelling the results will be high-level by design. If more detail is desired, then it will be necessary to move down the system architecture to the sub-systems and conduct analysis at that level. If still more detail is necessary, then equipment or components need to be analysed in the same way. At these levels the same principles apply regarding data normalization and the creation of Cost Estimating Relationships.

4. Compile Cost Estimating Relationships (CERs)

An excellent technique for creating Cost Estimating Relationships is step-wise regression. This is a technique used for choosing the variables (that is, cost drivers) to include in a multiple variable model. There are two types of step-wise regression: forward and backwards.

Forward step-wise regression starts with no variables and then adds them in controlled steps. At each step, the most statistically significant variable is added from the database until there are no parameters left worth adding. The choice of whether or not a variable should be added is made with the aid of the F-test, which determines the significance of the entire CER relationship.

Backward step-wise regression is the opposite, it starts with all the variables in the database assumed to be relevant to the model and removes the least significant terms until all the remaining terms are statistically significant.

An important assumption behind the method is that some input variables in a multiple regression do not have an important explanatory effect on the response. If this assumption is true, then it is a convenient simplification to keep only the statistically significant terms in the model.

One common problem in multiple regression analysis is multi-colinearity of the input variables, that is, input variables may be as correlated with each other as they are with the response. If this is the case, the presence of one input variable in the model may mask the effect of another input. A good example is shown in Figure 18.5 where PRICE KnowledgeManager has been used to select aircraft carrier systems from around the world. When the displacement (WT) is plotted against the Power (Max SHP) there is a close correlation; engineering instinct tells us that the heavier the ship, the more power is required to move it through the water. As a result, if one of these parameters is driving the cost model, there is no need for the other, it would be redundant – hence, multi-colinearity.

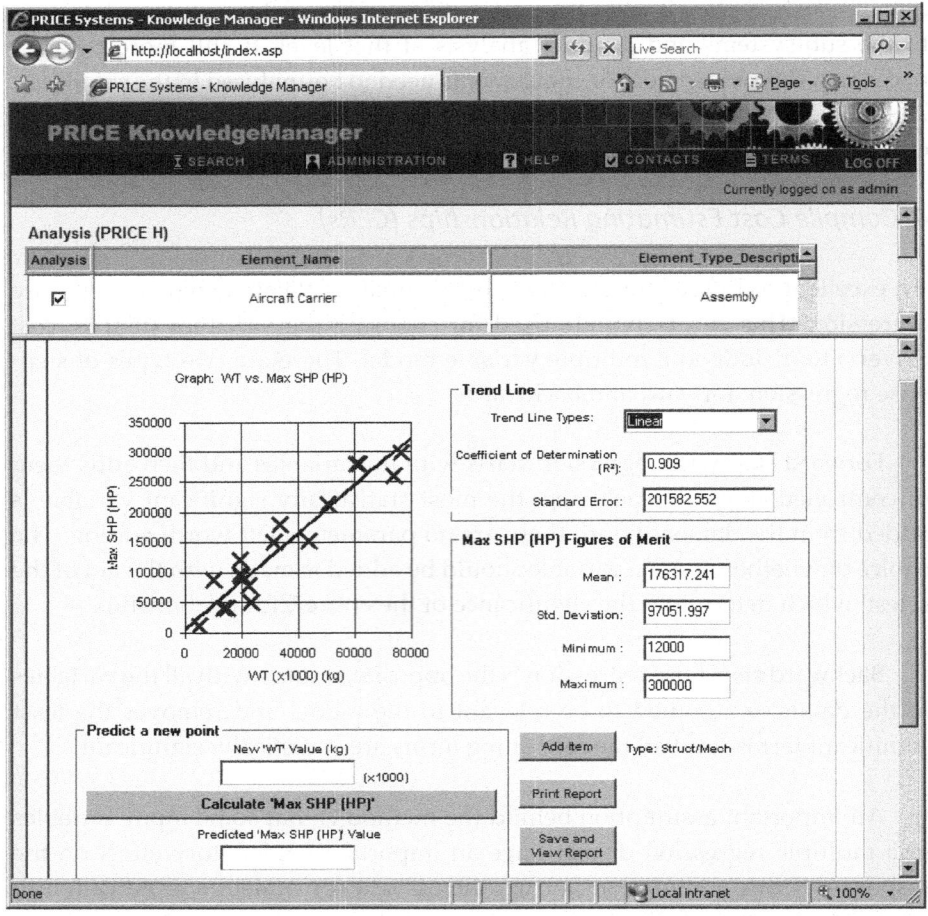

Figure 18.5 Multi-colinearity

A good cost model needs two things: a hypothesis and statistics to demonstrate its credibility. If a good statistical model is created, without a clear explanation of the model to peers or decision-makers, it will not offer convincing cost or schedule estimates. Likewise, if a good hypothesis is deduced (with an engineer's gut feeling that it is correct), but with no data or statistics to back it up, then the model will similarly not achieve widespread adoption.

After many false starts, this case study will be based on three key Cost Estimating Relationship hypotheses, based on:

1. weight and T_1 relationship;

2. technology improvement;

3. performance and design criteria.

Having normalized data and developed a hypothesis, it is now a question of finding a mathematical relationship for the first hypothesis. Using a statistical tool, it was possible to establish that a power form of equation was the best fit. This forms a non-linear relationship between weight and the cost of the first theoretical piece (T_1) as shown in Figure 18.6. In simple terms, the more the system weighs the more it will cost to produce.

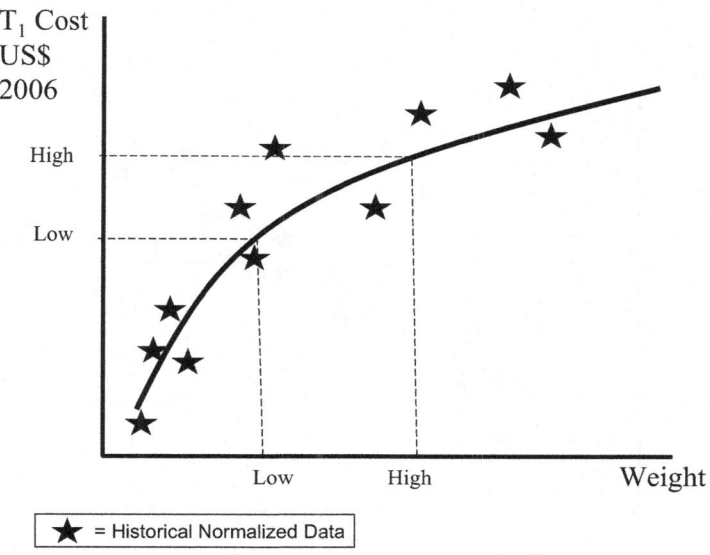

Figure 18.6   Weight equation determines T_1

The problem with a single hypothesis, like T_1, which is dependent on weight alone, is that life is unfortunately not that simple. If it were that easy, a large number of Cost Engineers would be redundant. This should not be disheartening, the predicted T_1 can be compared with the historical normalized T_1. In some cases the model is good, but in a lot of historical projects the prediction is a disaster. It is possible to consider other parameters from the historical database: look for the next hypothesis, the technology improvement.

This is where forward step-wise regression plays its part. If the difference between the predicted and historical T_1 values is calculated, it is possible to determine the influence on this difference, termed the Residual. This data has been sorted in chronological order using the In-Service Date (ISD) or Initial Operating Capability (IOC). The graph in Figure 18.7 shows the relationship between the Residual and the ISD. This is the result of reviewing many different parameters and applying engineering logic; it was not the first parameter to be considered. The graph indicates that there is an increase in the Residual as time passes. It is the influence of technology change, which weight alone cannot influence. As time passes, the technology becomes more expensive, accounting for increases in electronics and software. Using forward step-wise regression it is possible to intervene in the process of selecting an equation which best fits the data. With many statistical tools that use backward regression, the types of equations are fixed.

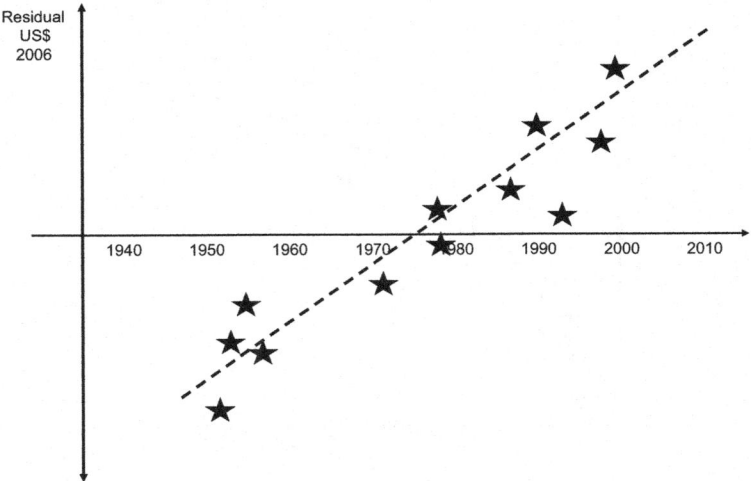

Figure 18.7 Residual (normalized history less prediction) versus In-Service Date

So, what has this achieved? Figure 18.8 shows the effect in simple terms. Building on the relationship that was already determined – weight and T_1 – it is now possible to influence the T_1 cost for technology. As can be seen in this figure, if a system is 4,000kg, the model was already able to calculate the T_1 cost. However, if this system was built at different periods in time, it is now possible to refine this estimate and either increase or decrease the T_1 cost depending on the date that the system is required in-service. This second influence (technology) is based on the Residual equation that was determined in the previous figure.

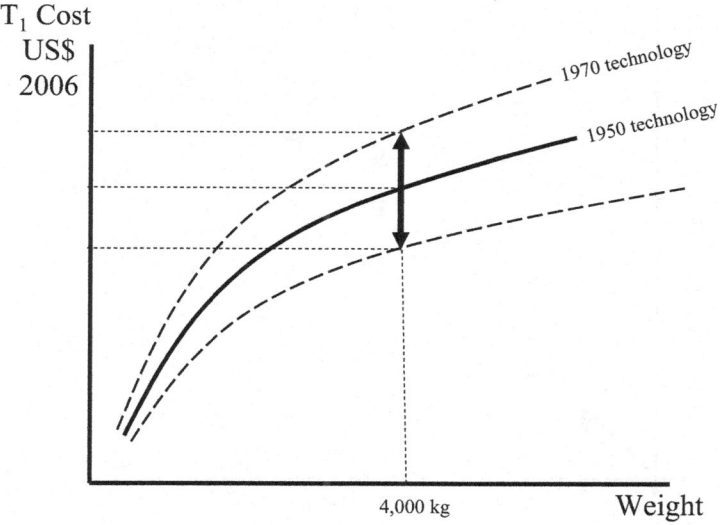

Figure 18.8 Considering technology impact on systems of the same weight

There is a need to size this technology Residual: it will not be a constant for all systems. For smaller systems it will be necessary to reduce the technology Residual and conversely for larger systems increase the technology Residual. It will not be possible to insert as much technology into a small system as a large system.

From here it is possible to repeat the step-wise regression. In other words, compare the combined weight plus technology prediction of T_1 cost with the historical T_1 cost and determine the level of accuracy of the model. If this needs to be refined, which can only be judged by the user regarding the level of accuracy required, the Residual is calculated again. It is then a case of determining the next performance or design parameter which has had an influence on the out-turn cost of the historical projects, for example, altitude, range, dive depth, speed, capacity, and so on.

Figure 18.9 provides a perfect example of this technique to demonstrate the methodology.

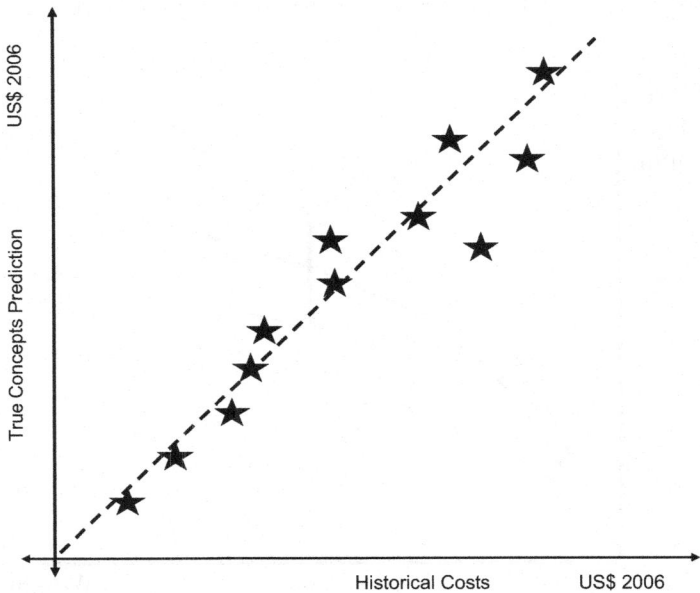

Figure 18.9   Combined effect of the three hypotheses

Being able to reliably calculate the production T_1 cost, by means of the third generation parametric framework, it is possible to estimate production costs:

- of different quantities using cost improvement curves;

- in different economics using escalation tables from the base of 2006;

- in different countries using Purchasing Power Parity (PPP) factors relative to the US dollar.

The same forward step-wise regression can be used on the schedule to predict the duration of development and production phases. Figure 18.10 shows the ability of a model to spread the cost over time using a beta distribution, once the cost and duration has been calculated. Again, functions like beta and uniform cost distributions are standard in a third generation parametric model and simple to implement.

Having discussed production CERs, it is important to consider other cost categories, for example non-recurring development costs. A three-pronged approach has been found most appropriate for non-recurring development costs, but first the hypotheses. Being able to calculate the T_1 reliably, it is assumed that the magnitude of the cost of the development phase will be relative to the magnitude of the production problem. In other words, the more costly the system is to produce, the more costly it will be to research, design, test and manage.

Figure 18.11 shows the approach that has been taken, dividing the non-recurring development costs into time-related, task-related and non-recurring manufacturing.

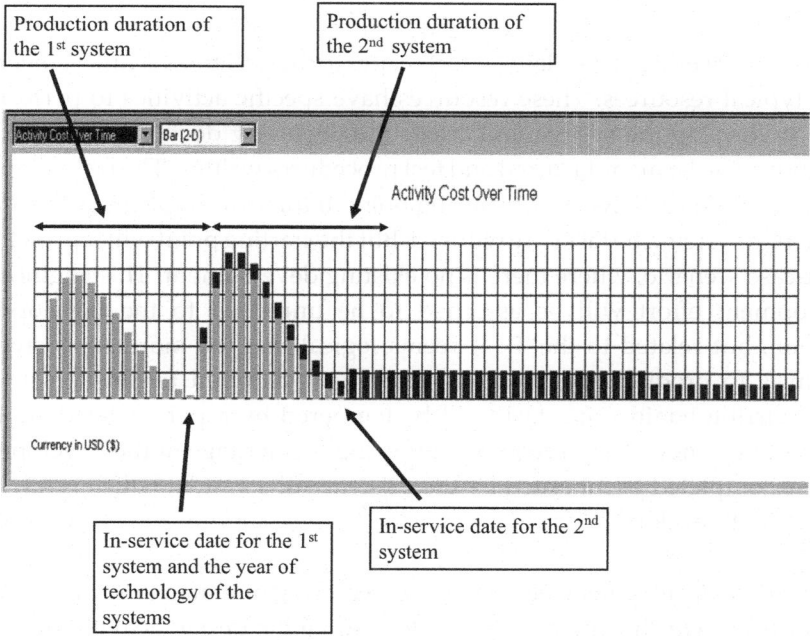

Figure 18.10 Demand and schedule

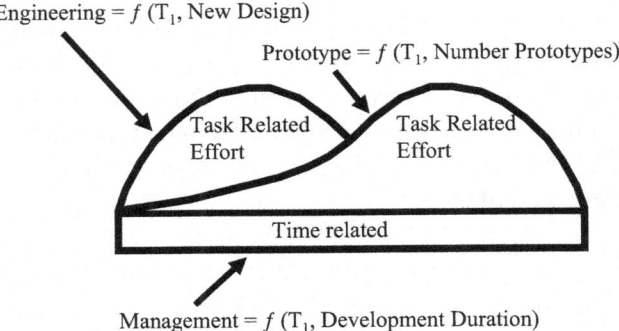

Figure 18.11 Non-recurring development costs

The time-related costs are management costs: they tend to exist from the start to the end of the development phase, only the magnitude (thickness) of the management task changes. Practically speaking, complex, expensive systems require one or more project managers, while a single project manager is well able to run several *cluster* projects if they are simple and cheap. Therefore, this time-related resource is a function of the T_1, to indicate the annual resource (or *thickness*) of the time-related activities, and the duration of the development program.

The engineering task-related effort relates to the design, system engineering, CAD typical resources. These resources have specific activities to perform in order to develop the system and create the necessary documentation so that prototypes can be manufactured and test procedures written. The tasks required will be modified relative to the starting point. If this is a simple modification to an existing system, the tasks are reduced. But the starting point will represent the worst-case scenario, where there is no existing development. This engineering development effort will be assumed to be unrelated to the development duration, but related to the T_1. It seems logical that the more expensive the system is to produce, the more engineering is required to develop it. This non-recurring engineering task will be tempered by a parameter describing the amount of new design required, inferring the heritage of the development already completed or inherited by the existence of a system which is having a new variant developed.

The final development element is non-recurring manufacture or prototypes, the hardware which is necessary to demonstrate the design actually fulfils the requirements. The Cost Driver here is naturally the number of prototypes required for the test program, but also the cost of each of these prototypes. Naturally the cost of each system can be related to the T_1. It follows that, if the production system is expensive, then the development systems will be expensive.

It now follows that these hypotheses need to be tested with normalized non-recurring cost data to establish their credibility. The important point to remember is that if I dismiss these hypotheses, do I have data and an alternative hypothesis to test? This is a high-level model for conceptual systems; the level of accuracy required must be kept in perspective.

The Operating and Support (O&S) cost, together with disposal costs, have not been forgotten and can be estimated. The O&S costs will depend on the usage of the system in the field annually. It will be best estimated as a fixed and variable calculation. The fixed costs like hangers, runways, docks and so forth will be a cost regardless of usage, but spares and maintenance costs will vary.

5. Input activities and resources

At last, the cost model can be constructed. The wonderful thing about using a third generation parametric model is that the model is completely visible to the Cost Engineer. It is possible to see into the model and observe the equations that have been produced. Figure 18.12 shows the transparent nature of the framework and some of the equations for the Military Helicopter Cost Object.

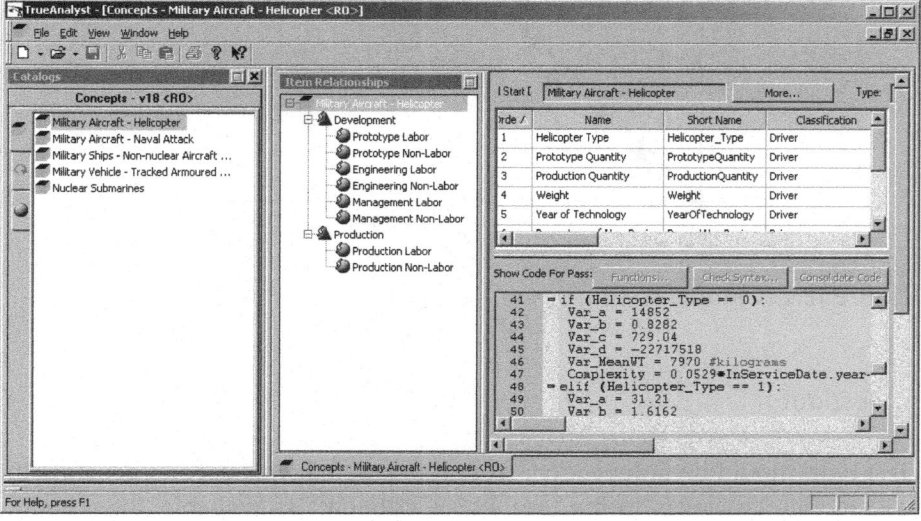

Figure 18.12 TrueAnalyst – equation visibility

6. Test and accept cost objects

Having constructed the catalogue, it just remains to test it. This should be simple as the open architecture of the third generation cost framework provides the quality assurance: if the model implemented in the cost framework provides the same answer as the cost research, then the model has been implemented correctly.

Figure 18.13 shows some testing conducted on the conceptual case study. The x-axis represents the historical data and the y-axis, the predicted cost from the model. If the model were perfect, the results would form a straight line. For this application (early conceptual studies) this type of accuracy (+/- 20 per cent) would be deemed acceptable. If more accuracy is required, then perhaps the system needs to be broken down into more sub-systems, or even equipment, and estimated using the more detailed hardware model.

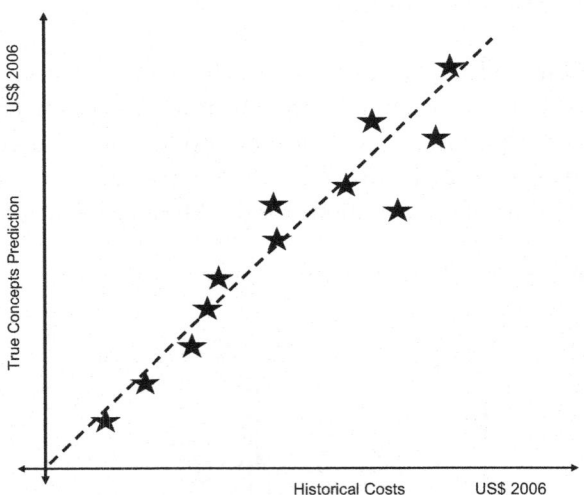

Figure 18.13   Model testing

7. Test data

Test data should be independent from the data that was used to generate the cost model. However, in the case where little data is available, there is little alternative to using the original data.

8. Validate and verify cost objects

Naturally, it is not just the figures that are important but also the distribution of the cost spread over time. By running the model with a variety of inputs and reviewing the outputs it should be possible to test and demonstrate that it is robust. Peer review will help continue the process of obtaining buy-in from colleagues: if these models are going to become widely adopted in the organization, every opportunity needs to be taken to expose them to scrutiny. With the client/server capability of a third generation parametric model, sharing these models will be easy and offers a return on the investment of building the models for the organization in the first place.

As a final point, it is also possible to provide validation and verification of the input data used to populate the Cost Drivers. This is just as important as considering the validation and verification of the Cost Estimating Relationships (CERs) in the model. If the CERs are good, but the input data is poor, then the results will be equally poor. Using a third generation parametric system it is possible to validate the input data in three ways (see Figure 18.14):

- *Input field* – ensuring the inputs are within the expected minimum and maximum of the parameter, for example percentages are above 0 per cent and less than 100 per cent.

- *Zone of tolerance* – when the model is run, it is possible to test the users' input to the Cost Drivers with non-cost design relationships and report the results. For example, it would be possible to test the feasibility of the design from the inputs which the Cost Engineer might enter for the number of aircraft and size of flight deck on an aircraft carrier. Providing the inputs are acceptable (that is, there are not too many aircraft on the ship that it becomes unstable), they can be used to estimate the cost. It is also possible to inform the Cost Engineer of the design's performance tolerance (for example, perhaps there is space for more aircraft).

- *Zone of intolerance* – if the non-cost design relationships determine that the inputs are unrealistic during the model execution, then they are deemed to be in the performance zone of intolerance. The third generation parametric engine will stop running and a suitable error message can be displayed (for example, you cannot have 100 aircraft on a 10,000-tonne aircraft carrier).

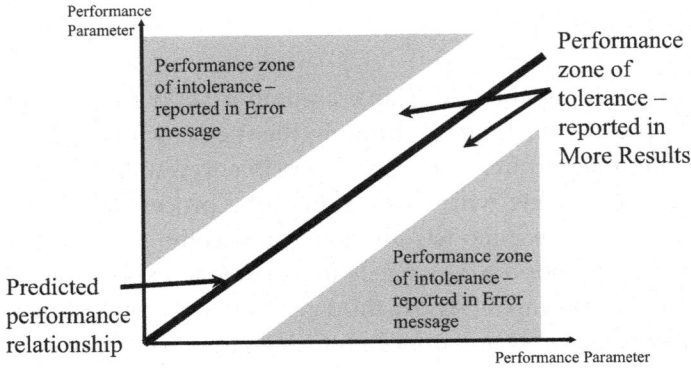

Figure 18.14 Non-cost performance validation

Summary

This chapter has reviewed the detailed steps necessary to design a cost model appropriate for implementation in the third generation modelling system. Using a case study of a conceptual model, it has been possible to demonstrate how this methodology can be used to build models that are statistically justified and may be explained to Cost Engineering peers and decision-makers.

This structured approach to model-building, combined with a commercially available framework has the following benefits:

- *simple cost models* – low number of inputs specific to the system – detailed architecture not required;

- *ease of use* – Graphical User Interface provides a common approach to inputs, risk analysis, outputs, escalation, capacity analysis, and so on;

- *design parameter checks* – early conceptual engineering principles can be applied;

- *open architecture* – the Cost Estimating Relationships (CERs) are visible and modifiable;

- *standard process* – a structured process for the development and acceptance of cost models.

19

How to ... Successfully Conduct Life-Cycle Costing

Parametric models exist for estimating the life-cycle cost of hardware, software and information technology. When it comes to estimating life-cycle costs, the challenge is not the cost modelling, but the data gathering. The acquisition life-cycle is defined as the procurement (development and production) plus the in-service (Operating and Support) phase.

This chapter introduces the concept of simple data gathering for Operating and Support (O&S) cost models. It was developed out of a recognized need by a cost model vendor: a need for information and data on Operation and Support costs. The challenge was not to develop a more sophisticated cost model, but simply to obtain the data necessary to run a cost model – detailed analysis is not a substitute for lack of data.

Methods of Data Gathering

The literature survey resulted in a number of different methods of data gathering, as listed below:

- *interrogating existing databases* – a review of the resources contained in databases in current projects and existing equipment databases;

- *questionnaires* – gathering information in a standard format using a form, either hardcopy (paper) or softcopy (electronically);

- *interviews* – using experts who have subject matter knowledge;

- *literature search* – searching through past studies, supplier documentation and other third party sources;

- *simulation* – it is possible to generate Operating and Support data from procurement data (development and production) with the use of mathematical relationships which have been researched based on previously gathered records. This is particularly useful when predicting the cost of future systems which are many years away from being produced, let alone operated in the field;

- *field tests* – the ultimate means of obtaining the Operating and Support data required by a model involving the measurement of repair and failure times in the field.

What is most important about the Operating and Support data that you gather? What are the attributes that distinguish between a good method and poor method of data gathering? Let us consider the criteria that can be used to evaluate the different methods.[1]

RESOURCE EFFICIENT

The most efficient data gathering method should not be resource intensive. In the modern economic climate, organizations cannot afford a huge team of staff devoted to data gathering. The methodology needs to be efficient in terms of both cost and schedule.

CURRENT

The data provided to the cost model needs to be current and up to date. It is of little interest that the data is plentiful and easy to gather if it relates to a prior technology that will not be employed. The data relating to diesel electric submarines can be of little use to Cost Engineers considering the cost of modern nuclear submarines.

ACCURATE

The data needs to be close to the original source and should be accurate. If other analysts have interpreted the information prior to you obtaining the data you

---

[1] Adapted from ACCA, The Certified diploma in Accounting and Finance, Management Accounting, Open Learning Workbook 1.

cannot be certain of its validity. The result of a cost model will only be as good as the information fed into it. If it is impossible to validate the data or to have confidence in it, then the cost estimate that results will be equally unreliable.

VALIDATED

The data needs to be validated and verified or reliable. It needs to be provided by a credible source which can be relied on. You would not ask a vehiicle breakdown an repair organization for the Operating and Support cost data related to aircraft.

COMPREHENSIBLE

Data is of little use if it is not understandable. A data dictionary is important to enable the person providing the data to have a complete understanding of the needs for which the data is being used, in terms of the data that they provide.

TIMELY

The Operating and Support cost data needs to be gathered in the timeframe available for the estimate. It is absolutely no use gathering the best data in the world, only to complete the exercise two months following the decision to procure a system that will not have the optimum life-cycle cost.

The sources of Operating and Support data are considerable; it should not be difficult to identify them in an organization. However, problems can sometimes arise due to internal politics, for example, when a project team is embarrassed about their statistics on reliability or cost. The list below provides an *aide-mémoire* for the sources of Operating and Support cost data:

- records from similar past projects;

- accounts and field data from similar in-service equipment;

- reliability and maintainability experts;

- suppliers' documentation;

- maintenance technicians;

- Integrated Logistics Support (ILS) Engineers;

- Failure Modes and Effects Criticality Analysis (FMECA);

- Logistic Support Analysis Record (LSAR).

Types of Data Required by Operating and Support Cost Models

When reviewing various Operating and Support cost models, it is possible to identify a trend in the types of cost data that these models require. In general terms they are split between parameters to describe the equipment to be operated and supported, the infrastructure to operate and support the equipment and a means of describing the Operating and Support policy.

- equipment description (for example, weight for transportation costs, volume for storage costs);

- corrective maintenance data (for example, Mean Time Between Failures [MTBF], Mean Time to Repair [MTTR]);

- scheduled maintenance data (for example, frequency of the maintenance);

- deployment data (1st line/organization, 2nd line/intermediate);

- maintenance policy or concept (contractor support or Government support);

- manpower (crew cost, training cost, support cost).

Advantages and Disadvantages of Different Data Gathering Techniques

Table 19.1 compares the different methods of data gathering. In terms of resource efficiency the simulated data is obviously the best. It requires little time and effort; however, although it might approximate to the correct data and need to be used in early concepts studies, it will be incorrect in most parts. Conversely, the interview, literature search and field tests will potentially consume large amounts of resources and require funding.

Table 19.1   Advantages and disadvantages of data gathering methods

✓✓✓ = Good, ✓✓ = Average, ✓ = Bad

	Resource efficient	Current	Accurate	Reliable	Understandable	Timely
Interrogate Existing Databases	✓✓	✓✓	✓✓	✓✓	✓	✓✓
Questionnaires	✓✓	✓✓	✓✓	✓✓	✓✓✓	✓✓✓
Interviews	✓	✓✓	✓	✓	✓✓✓	✓✓✓
Literature search	✓	✓✓	✓✓	✓✓	✓	✓
Simulated (Created by modelling)	✓✓✓	✓	✓	✓	✓✓	✓✓✓
Field Tests	✓	✓✓✓	✓✓✓	✓✓✓	✓✓✓	✓

The current nature of the data will best be served by the field tests. As tests are conducted of the equipment, they will provide an exact record of the equipment Operating and Support data. The simulated data is created from mathematical expressions. When these were researched and whether they are still current can only be answered by the suppliers of those routines. As described earlier, the use of simulated data can be extremely important when considering the Operating and Support cost of future systems. In this scenario there is no field data to gather, and the accuracy of the Operating and Support cost phases is perhaps less important to a current decision between design options. Operating and Support cost for the future system is many years away, providing an opportunity to refine the costs at a later time. The problem with interviews lies in the anecdotal nature of the data being presented. People tend to be optimists or pessimists; depending on whether they have had a good or bad experience with that piece of equipment, different maintainers will provide different answers. The best accuracy is going to be achieved by measuring the necessary parameters in the field.

Data reliability is low for interviews and simulation and high for field testing for the reason described above. But understanding the data introduces a range of new influences. A good understanding of the data is essential; a data dictionary is a good starting point for defining the immediate boundary of the data. As such, existing databases and literature surveys are not a particularly good technique for obtaining data: someone else will have gathered the data and will have made their own assumptions in the process. Indeed, if the data has been processed, it is not (by definition) raw data. Conversely, questionnaires, interviews and field tests will provide an opportunity to clarify the boundary, assumptions or definitions of the data being gathered.

Finally, the field test and literature survey are not likely to produce results within the necessary timeframe. If the Operating and Support cost is required to influence a decision in the near term, the best methods would be the simulation, interview and questionnaire to ensure the data is available when it is needed.

Taking all of these criteria into consideration, the technique that has the fewest drawbacks whilst providing greater than average performance is the questionnaire.

Life-Cycle Questionnaire

It is important to distinguish the roles involved when designing a questionnaire. There will be a group of users who are interested in the gathering of the data. This group requires the data to populate a cost model; we will label them the *Customers*. This group is looking for ways to make their budget stretch over more programs.

There will also be the organizations that are responsible for providing the Operating and Support data. These are staff who will not benefit directly from the data gathering, therefore, their task needs to be as undemanding as possible to enable them to provide the Operating and Support data. We will label this organization the *Suppliers*.

Looking more closely, two types of Suppliers can be identified. Those that reside within the Customer's organization and those that will probably lie outside the management structure of the Customer organization. Examples of the external suppliers include:

- subcontractors;

- equipment suppliers;

- support organizations;
- current legacy projects.

Examples of internal suppliers include:

- project teams;
- Integrated Logistic Support (ILS) departments;
- Integrated Project Team (IPT) staff;
- future projects departments.

The example Life-Cycle Questionnaire that will be used here has been developed by PRICE Systems and was released in the PRICE Estimating Suite 2003 it demonstrates some good practices for an O&S questionnaire. It is a Microsoft Excel workbook that contains several worksheets. In summary the various worksheets are self explanatory:

- Instructions for Customer;
- Instructions for Suppliers;
- Example;
- Questionnaire;
- Builder.

The purposes of the different sheets are contained in the following sections.

'INSTRUCTIONS FOR CUSTOMER' SHEET

This sheet contains different embedded file formats (including rich text format and ASCII Text format) of the same document. This ensures that most people will be able to access the files even if they do not use a Microsoft product for their word processor. It is important when creating a questionnaire that it is as user friendly as possible. Any difficulty, like problems opening a file, can be used as an excuse to return the document unpopulated.

These files are provided to give guidance on how the Customer will use this Life-Cycle Questionnaire. Following a brief introduction the topics covered in the customer guidance include sections on:

- how to prepare a questionnaire to be sent to the supplier;

- what equipment deployment data is required;

- what needs to be done when the supplier populated questionnaire is returned.

'INSTRUCTIONS FOR SUPPLIERS' SHEET

This sheet again contains different file formats (including rich text format and ASCII Text format) of the same document. Each file is identical and is provided to assist the Supplier in completing the Life-Cycle Questionnaire. In this case the topics covered include:

- introduction providing explanation of why the data in the questionnaire is requested and what will happen to it;

- the deployment data regarding the project;

- how to create an Estimating Breakdown Structure (EBS);

- what Operating and Support costs are needed and their definitions;

- overview of the cost model which will be employed to calculate the costs when the questionnaire is returned.

'EXAMPLE' SHEET

The Example Sheet contains a set of data for a 'GPS Receiver system'. This is an example of a completed questionnaire sheet. As can be seen in Figure 19.1, this example can be used to guide the Supplier through the process of completing the questionnaire – the type of inputs required and the detail desired.

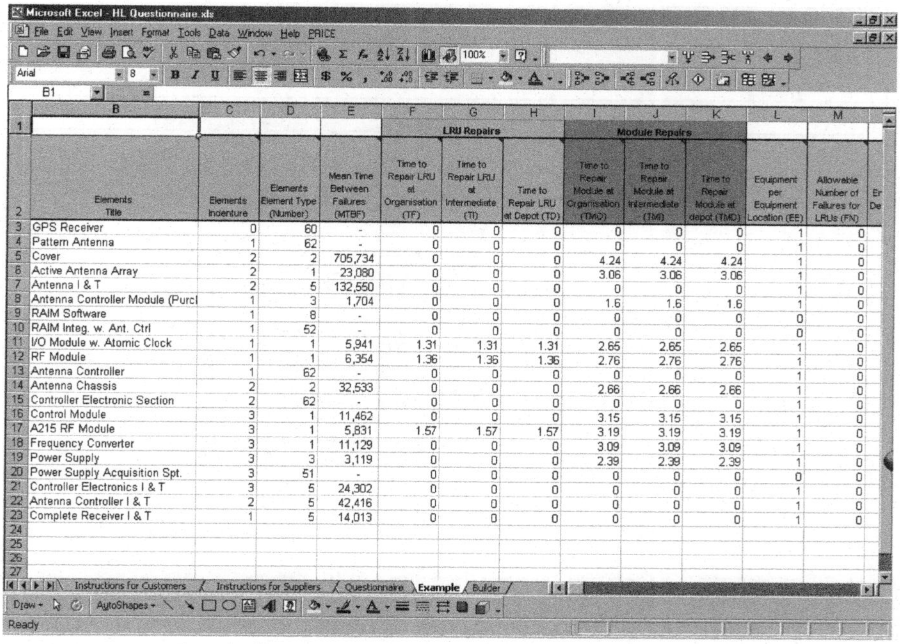

HOW TO ... SUCCESSFULLY CONDUCT LIFE-CYCLE COSTING

Figure 19.1 The example questionnaire

'QUESTIONNAIRE' SHEET

The sheet in Figure 19.2 is the Life-Cycle Questionnaire itself. It contains the hardware parameters that need to be populated in order to run the parametric cost model. Each column has a header that provides a brief description of the parameter including its units (hours, currency and so on) and its parameter abbreviation (MTBF, DSTART, and so on).

Each row needs to be completed for each element in the Product Breakdown Structure (PBS). This will be specific to the Supplier's product or equipment.

The column headers contain comments that perform the function of a data dictionary. They define the interpretation of the input for the Supplier. As can be seen, the Mean Time Between Failures (MTBF) has been defined and exemplified in the screenshot.

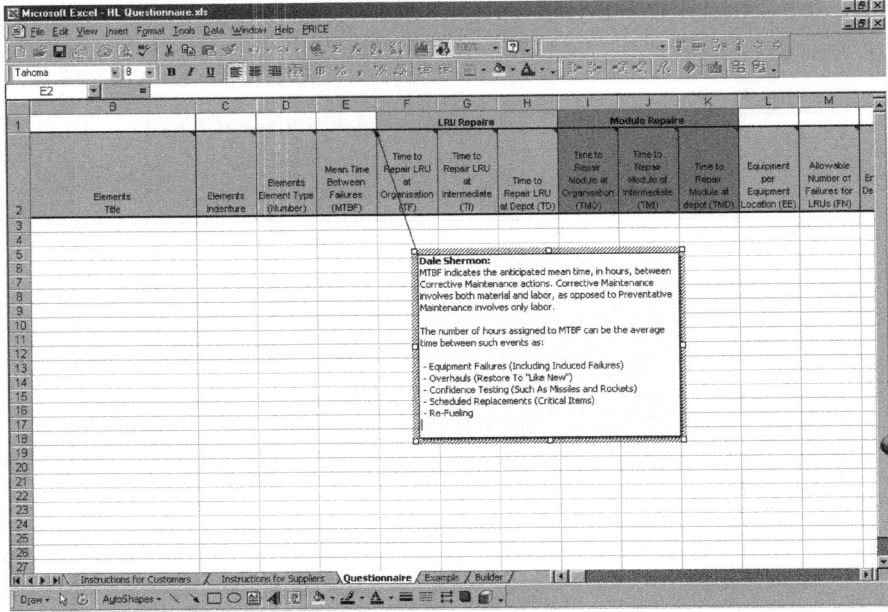

Figure 19.2 The questionnaire

COST MODEL INTERFACE – 'BUILDER'

A simple mechanism is needed to transfer the questionnaire data to any parametric cost model. In this example a Builder sheet is the key to creating a parametric model Product Breakdown Structure (PBS). This sheet has the ability to interface with this particular cost model. As the Supplier will not necessarily have the parametric model, this Builder sheet will not be provided to them.

The Data Gathering Process

The questionnaire is relatively simple once the roles of the Supplier and the Customers have been identified. Figure 19.3 provides a summary of the process and the responsibilities of the Customer and Supplier.

To start the process the Customer must have a good understanding of the system that is to be subjected to life-cycle cost estimating: the system architecture, the contractual work share and make/buy decisions of the equipment that constitute the system. Once determined, these provide the key system attributes. The Customer is in a position to prepare a Supplier's Questionnaire: the Life-Cycle Questionnaire, Example and Suppliers Instructions.

Two parallel activities then take place. The Customer needs to continue preparing the information around Deployment and Employment and the maintenance concepts. The Supplier is probably not able to contribute to the Deployment and Employment decision part of this process as the final recipient of the cost estimate and the Customer have that contractual relationship. The Supplier may have opinions on the most effective way to support and maintain their equipment, but the ultimate Customer may be constrained by a fixed or restricted capability for support and maintenance.

Meanwhile the Supplier, either internal or external, may populate the questionnaire. They have been identified as the best source of data, but may only have part of the information required. For example, the maintenance data may derive from one source and the spares data from another. Alternatively, some sub-systems may come from one Supplier (possibly internal – 'Make') and other sub-systems might come from another separate Supplier (possibly external – 'Buy'). Eventually, the completed Supplier Questionnaire, containing the Operating and Support data, will be collected by the Customer and used in the Operating and Support cost model.

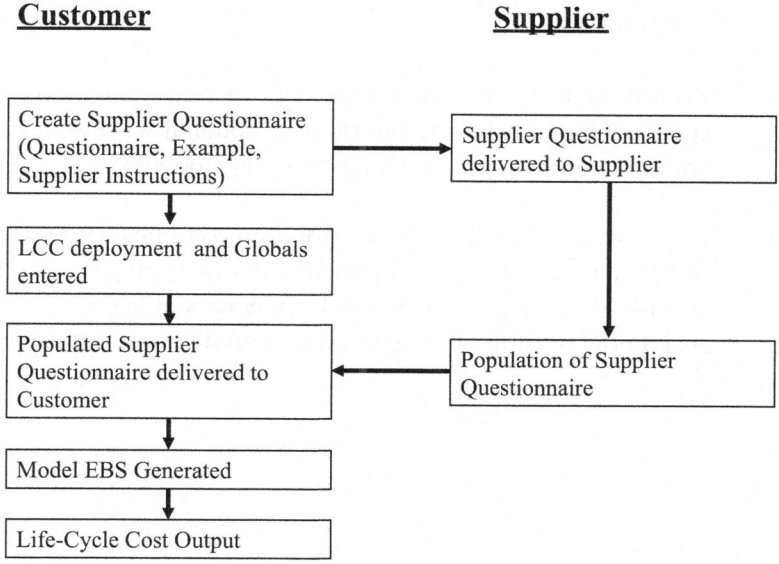

Figure 19.3   Questionnaire process

The questionnaire can also be used to acquire data for a future system through a process of structured interviews, the questionnaire providing prompts for a meeting with the supplier of the data.

Finally, the Customer uses the parametric cost model through the spreadsheet to generate the Product Breakdown Structure (PBS) automatically. Providing the Global and Deployment information have been completed in the model by the Customer, the System Life-Cycle Cost can be calculated.

Summary

In this chapter you have learnt about the application of a questionnaire to the Operating and Support input parameters for a cost model. These lessons equally apply to the other parametric models:

- *Consistent methodology for data gathering* – A questionnaire provides a method of gathering Operating and Support cost data that complies with the criteria that were stipulated.

- *Simple interface for non-parametric trained users* – The questionnaire, being a Microsoft Excel solution, explains the model parameters such that non-LCC trained engineers can understand and gather the data.

- *Accelerated process for data transfer* – If the questionnaire is used in conjunction with the parametric cost model, it will result in speed and ease of Operating and Support costs modelling.

- *Improved quality of data* – As the solution complies with the criteria of the data gathering requirements, it will result in more timely, consistent, accurate, current and understandable data to feed the cost model resulting in improved cost predictions and lower LCC.

20

How to ... Accomplish Knowledge Retention

Retaining corporate knowledge is essential in any organization that has ambitions for growth.

If a cost estimate or a price is being presented to senior management or a customer, unless it can be substantiated the figure it is, at best, an opinion. To have a credible cost estimate needs a strong Basis of Estimate (BOE). Knowledge management enables Cost Engineers to provide credible estimates quickly and accurately based on the historical information readily available to them. The BOE provides decision-makers with the confidence that the estimates are based on sound, auditable foundations, which ultimately assists companies in making faster, profitable decisions. If you have confidence in your costs you are more likely to bid at the winning price.

Parametric cost model vendors have recognized the needs of the Cost Engineer for knowledge management and developed the requirements further specifically for parametrics:

- to share parametric cost data across teams or functions;

- as a better way to store data;

- to offer easy access to the data;

- to ensure consistent, logical organization for rapid searches and analysis;

- to work on the basis of easily maintainable industry standard software technologies;

- which can exploit internal historical data.

The most accurate Cost Estimating Relationships (CERs) are based on an organization's past experiences. Increased confidence in estimates is based on repeatable and reliable metrics. Parametric model vendors, are trusted suppliers of historical, industry normalized data points to complement and supplement an organization's own metrics or knowledge.

Knowledge Management

A good knowledge management system provides a common, unified storage, retrieval and analysis centre for software and hardware estimating data, but also non-parametric performance characteristic data (for example, speed, power, torque, acceleration, pressure, manufacturing process) and Cost Driver data (for example, component count, external finish, In-Service Date). It also provides estimate-tracking audit trail data (who did what and when) and statistical analysis data.

The simplest use of the stored historical knowledge is obviously estimating by analogy. If a new system to be estimated is similar to a system retrieved from the knowledge management system, this provides a good starting point. Keywords will be used in the knowledge base to aid the retrieval of the system. This will facilitate structured searches of the knowledge base.

How to Achieve Knowledge Retention

The following is intended to be a guide to assist trained users of parametric models with the introduction and implementation of knowledge management into their organization. It provides hints concerning general points to be considered, with the introduction of knowledge management reflecting the experience of consulting staff.

STAGE 1: DISCUSSION

Initial discussions should be held with senior management to confirm the program of work, identify the task leaders for the data to be collected, and in particular to confirm the scope of work and deliverables. It is good project management practice to ensure that everyone is aware of the objectives of the task to ensure that nobody is surprised by the result on completion.

This discussion should result in the construction of a table showing the equipment produced by the organization, with identification of old and new products. This will then form the basis of a knowledge database data-gathering program. It is critical that time and resources are not wasted gathering data when, at the *completion* of the program, the data in never used again.

STAGE 2: DATA GATHERING

The boundary and assumptions related to the information that will populate the knowledge base need to be determined. These must be agreed with fellow Cost Engineers. It is possible that many of these working practices have already been documented and agreed, for example the global values used in the model. Other questions to consider include:

- Are metric or imperial measurements being used?

- What are the escalation table defaults, Consumer Price Index (CPI) or Retail Price Index (RPI)?

- What are the General and Administrative (G&A) fee, overheads, profit and fee rates?

- How are electronics and structure differentiated?

- What version of the parametric model is being run?

STAGE 3: NOTE TAKING

When assumptions are made or there is uncertainty regarding any aspect of the data or project, this needs to be recorded. To enable the users to make sense of the information when retrieving the data from the knowledge base, they need to be armed with any extra data obtained during data gathering. The best way to achieve this is by using the notepad facility within the models as information entered here will also be captured in the knowledge base.

STAGE 4: DEFINING USERS AND GROUPS

When the knowledge management software is initially populated, it can typically be divided into sections called *groups*. These groups are used to segment the data stored in a single knowledge base, which is preferable to having duplicate knowledge bases leading to a situation where nobody is able to identify the master knowledge base. Users, whose access to the system is password protected, are given access to the groups by the system administrator. The groups and users are defined by the system administrator who should have flexibility to tailor the knowledge management systems to the optimum configuration giving access to the most relevant users for future program estimate without compromising commercial confidentiality or security classifications (like secret or restricted historical data).

The two examples in Figure 20.1 show different configurations, one determined by the functionality of the estimate (Army, Naval, Air) and the other by estimate type (bid/proposals, calibration of internal items, calibration of suppliers cost). Admin has access to all groups and other users are given authority to access various groups outside their own, for example, Person C can only view the Army or Supplier Assessment data in the respective scenarios.

STAGE 5: DATA MANAGER

Typically, a cost research organization employs a Chief Scientist to lead the cost research team. Amongst his or her responsibilities is the validation and verification of the data that appears in the knowledge base. The quality of the data and its sources are important to cost research organizations and this is therefore an important role.

When starting to populate a knowledge management system with data, it is wise to consider appointing a Data Manager or Knowledge Manager to be the single point of focus for all data. This does not mean that he or she is required to produce all the calibration or data files, but simply that he or she has a scrutiny role to ensure that the database does not become populated with misleading or inaccurate data and conforms to the consistent approach that the organization has agreed for data gathering.

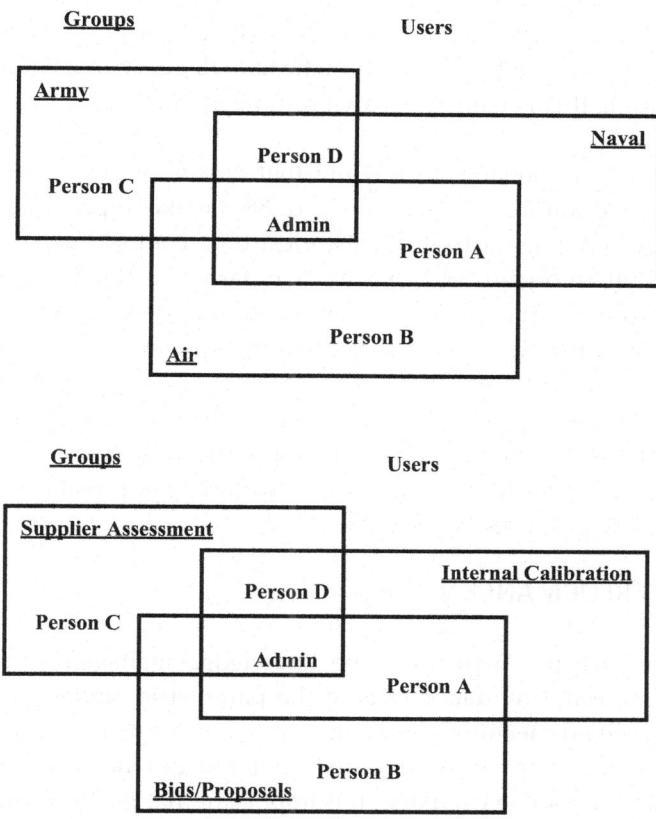

Figure 20.1 Security

STAGE 6: KEYWORDS

Hierarchical and general keywords need to be agreed upon. To enable the organization to obtain maximum return on its investment, it is important that knowledge can be retrieved quickly and logically.

Parametric estimates are based on a Product Breakdown Structure (PBS), it therefore makes sense to save calibrated items in a form that mirrors this structure: if in the future the organization wants to perform an estimate by analogy, the whole PBS for a past program will need to be retrieved.

General keywords are useful for retrieving data across several PBS. For example, if there is a need to analyse aluminium alloy castings to determine an appropriate Manufacturing Complexity, the keywords 'aluminium', 'alloy' and 'casting' would need to be stored in the knowledge management system to facilitate finding them again in a search routine in the future.

It would be reasonable to assume that customer standard breakdown structures (for example, US DoD Mil. Std. 881) make appropriate guides for sourcing keywords against which historical data could be stored. However, it could be that an organization has its own particular standard that needs to be adopted to be compatible with other business systems like the accounting, finance, manufacturing, EVM or project control systems.

Keywords also provide continuity. They ensure that the term 'Reaction Control Systems' is entered and retained as such and not as 'RCS' or 'React. Cont. Sys.'. If all three terms were used, it would be impossible to search and find all possible instances of this data.

STAGE 7: PERFORMANCE VALUES

The performance parameters for the knowledge management system will need to be agreed. The data sheets in the parametric model typically allow customer-specified meanings for capturing text or numerical values. As such they enable any performance values to be entered, but inconsistent values will mean the data cannot be validated. It is important that the performance values are consistently applied. Examples of performance values include:

- power – horse power or kilowatts;

- velocity – miles per hour, kilometres per hour or metres per second;

- range – metres or kilometres, feet or miles;

- altitude – metres or kilometres, feet or miles;

- diving depth – metres or kilometres, feet or miles.

Each system type requires appropriate, agreed parameters including the magnitude of the units to be applied. This should be a common task applied across all data sets.

To ensure that the analysis and parameters in the pick lists make sense, use titles to define the data parameters rather than abbreviations such as 'Val 1', 'Val 2', 'Val 3'.

STAGE 8: NON-PERFORMANCE VALUES

The same rules apply here as for performance values. The Cost Engineering team in an organization needs to agree on the format or definition of the values that are going to be used. For example:

- type of estimate – ROM estimate, actual, detailed estimate;

- data source – research, public domain or publication, quotation;

- manufacturer;

- country of origin;

- In-Service Date;

- identification of the engineering and manufacturing processes – CAD tools, NC machining, high-speed machining, casting, forging, and so on.

This should allow the selection of these criteria in subsequent queries.

STAGE 9: SUITABILITY OF DATA

Part of the process requires establishing the suitability of data to be stored in the knowledge base. The originator of the data should be contacted to evaluate its suitability for use in a knowledge base. The future use of any data also needs to be taken into consideration. Particular attention needs to be given to the consideration of missing data. If there are gaps in the data set, it is possible that an exercise to gather the missing data could be considered. Reasons for a data gathering exercise will need to be explained to senior management in order to determine if such an exercise would represent value for money.

STAGE 10: DATA CALIBRATION AND STORAGE

Time should be spent reviewing the complexity of each of the systems and storing the performance data in the data sheets within the model. Finally the parametric data should be transferred to the knowledge base.

If any of the mandatory data required within the parametric model is missing, the data item could be included, uncalibrated, in the calibration file, so that if the information becomes available in the future, these data items can easily be added to the knowledge base.

It will be important to validate the data prior to the data being stored and released for general use. Ideally one person, the Data Manager or Chief Scientist, should be responsible for data storage following an independent review of the data. This ensures that data integrity is maintained and the knowledge captured is useful to Cost Engineers in the future.

STAGE 11: FINAL REPORT AND PRESENTATION OF THE RESULTS

The knowledge management task starts with a discussion, to complete the task a report needs to be written explaining the results of the calibration and the analysis that can be performed using the data and the knowledge management system. Preparing a presentation around the report will help ensure buy-in from the Cost Engineering team.

FINAL THOUGHTS: TIMING

Capturing historic data is never a popular task. However, today's information is tomorrow's historic data. So rather than spending time in the organization's archives trying to find information or people who can remember anything about a program, it is better to start with today's data: it will be surprising how much knowledge will have been gathered by this time next year.

Case Study

This case study demonstrates the features expected in a knowledge management system. It involves trend analysis performed against characteristics related to cost. In the defence environment, once the threat has been identified, Operational Analysis (OA) can be used to determine the systems most suitable to counter that threat; an example could be the need for an aircraft carrier as part of a force consisting of a mix of other military assets. Operational Analysis will determine the optimum number of aircraft needed for any deployment, and the time to deploy them will establish system characteristics such as size of the flight deck. From these facts it is possible to calculate the displacement of the ship.

A knowledge management system can been used to retrieve data related to all the aircraft carriers that have previously been stored. Figure 20.2 is an example of such a knowledge management system – in this instance the web-based systems shows that 30 carriers have been found; their associated data is visible including the escalation and financial data stored with the estimating element.

It is now possible to perform a regression analysis on the same data set to compare the In-Service Date to the Manufacturing Complexity of all the aircraft carriers in the same manner. This Manufacturing Complexity parameter (see Figure 20.3) is one of the variables used in the model. Ideally it would be possible to add this item to the equipment breakdown structure automatically.

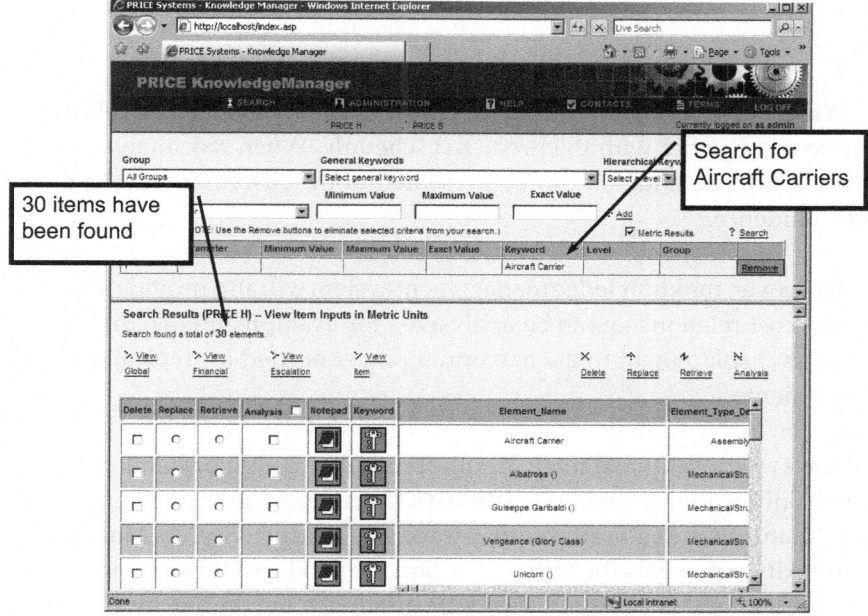

Figure 20.2   Aircraft carriers retrieved

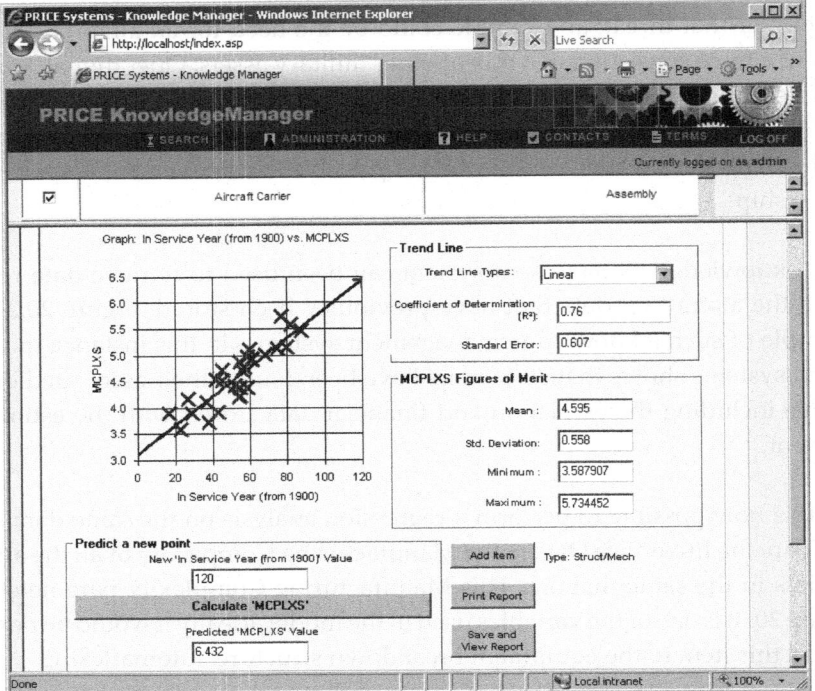

Figure 20.3 Analysis of aircraft carriers

When other variables are added, the program costs can be established with confidence together with the predicted schedule. When risk analysis has been added, a full presentation can be presented to the customer with justification and confidence.

Moreover, the knowledge management system will also provide information for non-cost relationships to be analysed – for example, it is possible to relate the ship's displacement to the maximum power needed for typical operational requirements.

The system can have many applications: space, military and commercial, limited only by the imagination of the user. Consider, for example, benchmarking your organization against the industry average productivity? As Manufacturing Complexity represents the technology and the productivity of an organization, if the technology is constant, then any difference in complexity is a result of efficiency or inefficiency. Ideally a knowledge management system containing thousands of elements from internal and external sources which will be easily accessible when required. Although providing a good starting point for analogous estimates, it also provides a rich source of benchmarking data. Once an organization has calibrated itself, it is possible to compare its efficiency against the industrial norm stored in the knowledge management system for identical technologies.

Summary

Due to the computerized nature of parametric cost models and the formality that they bring to the estimating process, they are first rate systems for implementing knowledge retention in an organization. In this chapter you have learnt that parametric models have knowledge management capabilities built into them and can aid the organization of past program cost and schedule data for future usage in bids.

21
How to ... Present the Results

A critical part of any estimate is the presentation of the results in a manner that the customer or decision-maker is able to interpret easily and quickly. There is little point is spending huge amounts of time refining the estimate if it is too late to influence the decision or it is presented in a way that is too complicated.

The first step must be to review the estimating request and determine what the original 'exam question' was. It is possible that the program has evolved and the original question asked is no longer relevant. A wise Cost Engineer once told me: 'There is never enough time to do the cost estimate properly, but there is always enough time to do it twice.'

The next step is to put the estimate into context. It is too easy to fall into the trap of giving the customer or decision-maker the total estimated cost. They will want the answer, but it needs to be put into perspective. The most widely recognized means of achieving this is with an assumptions document of some kind. There are several formats and document types, but in the UK Ministry of Defence (MoD) this is recognized as a Master Data and Assumptions List or MDAL. In the US Department of Defense (DoD) the document is called a Cost Analysis Requirement Description or CARD. The purpose of the document is to place a boundary around the estimate – record all the inclusions and exclusions – also to register the data sources to enable others to scrutinize them and suggest other places where suitable data could be obtained.

Finally the estimate needs to be presented – as a cost report, vetting document or a presentation. Putting the cost and schedule estimate in context is absolutely necessary. There is no better way to provoke an argument than by providing an estimate without the necessary introduction of the methodology, assumptions and data sources.

Typically a cost report or vetting template is used to conclude the costing exercise. This will put the estimated cost into context. It should ideally have the following contents:

1. Executive Summary
2. Table of Contents
3. Distribution
4. Glossary
5. Introduction
    - Background
    - Previous Estimates
6. Data Types Utilized
7. Estimating Methodology
8. Data Sources
9. Assumptions
    - General Assumptions
    - Financial Assumptions
    - Program Assumptions
    - Exclusions from this Study
10. Results
11. Study Results
12. Discussion of Results
13. Future Considerations
14. Recommendations
15. References

One of the most useful indicators for a decision-maker is the derivation of the estimate: how the total estimate is broken down into different types of estimating methodology or data source – for example, to what degree have parametrics been used, what proportion involves detailed engineering estimates, or how much is based upon supplier quotations. A simple pie chart (see Figure 21.1) will help a decision-maker take confidence in the Cost Engineer's ability to prepare an estimate.

One of the more convincing reports to present to a senior manager is a sensitivity analysis. This can be attained by taking the Cost Drivers for the estimate one at a time and varying the input parameter value, then recording the effect that the change has on the cost output – relative to a baseline cost – and finally then restoring the Cost Driver to its original value and changing the next Cost Driver to determine its sensitivity to the output cost. In many cases it will be determined that inputs have very little effect, in other cases the Cost Driver may have a significant impact on the item itself and other items in the overall Cost Breakdown Structure.

It is also highly likely that there is a non-cost aspect to the question that was originally posed. Costs are usually linked to investment, which should result in a benefit. The benefit is likely to be the product of another process such as Operational Analysis (OA), Benefit Assessment, Business Resilience Index, and so on. These non-cost aspects are combined with the costs to determine the benefit outcome that will be the result of the expenditure.

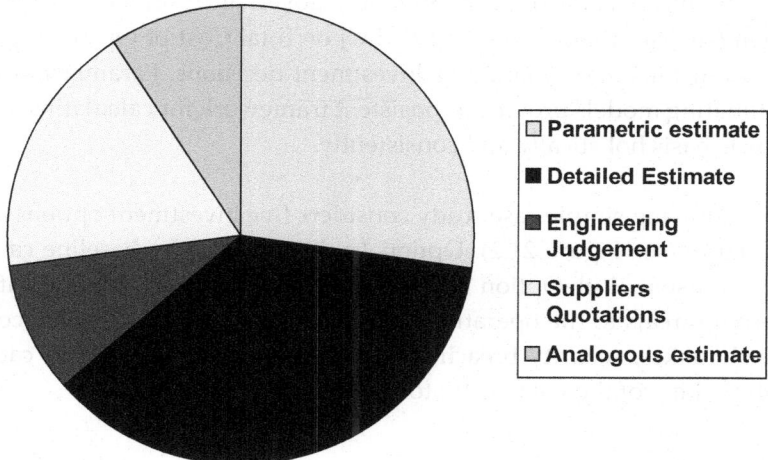

Figure 21.1 Derivation of the estimate

Business Resiliency Case Study

Every large commercial enterprise must deal successfully with challenges of planning, coordinating and managing information technology (IT) governance and compliance. In most organizations, however, IT decision-makers lack the oversight they need into spending, performance and Whole Life Cost (WLC) or Total Cost of Ownership (TCO) of these essential programs.

IT project decision-making continues to be based on reactive responses to presumed needs, inadequate risk-versus-value assessments, and over-optimistic projections of business outcomes. Whatever business case is made to launch a governance or compliance initiative often occurs at a departmental or divisional level, it rarely includes the necessary factual details to ensure accurate estimates, and increases IT spending without consideration of overall enterprise objectives or priorities. Employing non-biased performance measures, benchmarks and models reveals the true risks and cost benefits of IT governance and compliance programs.

Many initiatives in the banking and insurance industry have pointed to the need for understanding, measuring and strengthening Business Resiliency. However, there is a minimum cost required to achieve a level of resiliency: what is desirable is an optimized resiliency effectiveness for each dollar spent as measured by the Resiliency Index.

Frameworks such as CobiT 4.0 and Val-IT align IT strategy to business strategy within a context of compliance, governance and operational risk management. The Whole Life Cost (WLC) or Total Cost of Ownership (TCO) provides a means of evaluating IT investment decisions. Parametric software cost estimating models provide a consistent framework for calculating WLC on a life-cycle basis holistically and consistently.

The following simple case study considers five investment options around a data centre (see Figure 21.2). Option 1 will consider the baseline case with business as usual. With Option 1 there will be no consideration of the influence of external threats to the operations of the business. External events could be fire, earthquakes, security breaches or any other event that could cause the normal working of the data centre to be disrupted.

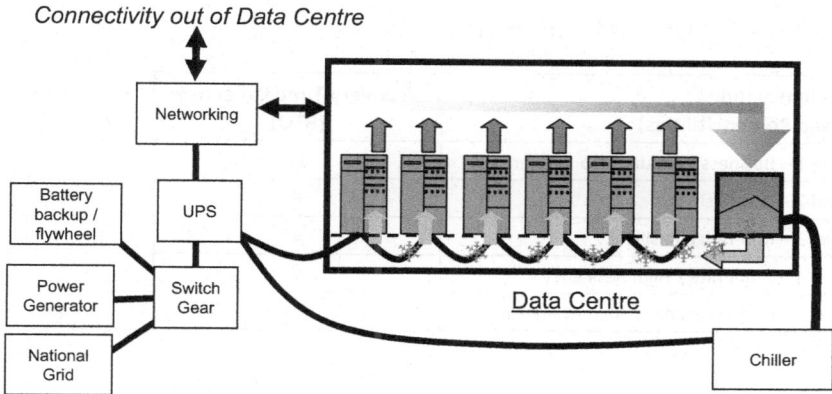

Figure 21.2 The data centre problem

At the other extreme, Option 5 will consider high resiliency of the business with a resulting high availability of the data centre capability. Various studies will be conducted, mitigation plans considered and implemented for:

- seismic safety;
- fire suppression;
- security;
- monitoring and control.

In the case of the data centre, the WLC can be estimated and the financial implications assessed. The resiliency of the data centre can also be established and it is possible to calculate the resiliency in terms of the Recovery Time Objective (RTO) in minutes. This represents the anticipated time to bring the data centre back online and functioning after an event of some nature. The resiliency represents the time the data centre is lost to the organization.

The results are presented in tabular form. Table 21.1 displays the results of the resiliency study and shows the Recovery Time Objective as the measure of resiliency. However, when presented in a table format these are merely cold hard facts and therefore not easy to make decisions upon or interpret.

Table 21.1 Data centre problem

Resiliency Index (Fixed Power Utilities)	Recovery Time Objective (RTO)	Servers/Racks
Level 1 – Business as Usual – No Resiliency	12	40
Level 2 – Low Resiliency	6	48
Level 3 – Mid-range Resiliency	4	52
Level 4 – Moderate / High Resiliency	2	60
Level 5 – High Resiliency / High Availability	1	80

In this example the best presentation of the results would be the combination of the WLC and the RTO in a graphical format. This cost-versus-benefit presentation provides an easy visual representation of the investment to be made and the likely benefit that would result in such an investment.

Ideally, the cost-versus-benefit graph (Figure 21.3) would identify the objective and threshold for the cost and the benefit. This provides no-go zones identifying areas within which options will not be considered. It also helps eliminate some of the levels in our case study and focuses on the options that represent conceivable, affordable solutions.

Despite the usefulness of the graph it can still be difficult to decide on the most cost-effective solution. It might ultimately be sensible to select options that

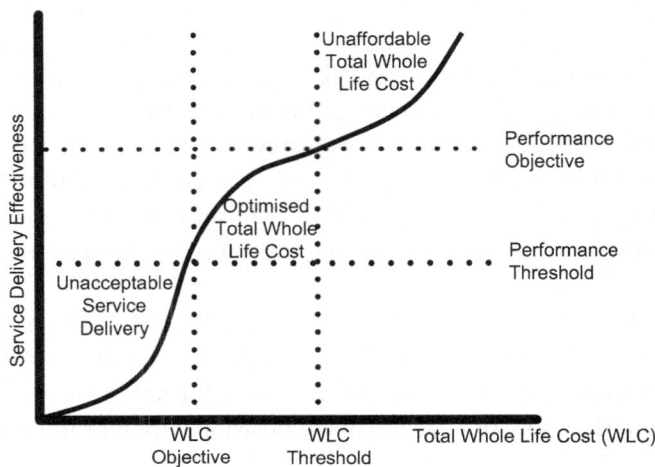

Figure 21.3 Graphical presentation of cost-versus-benefit analysis

provide either constant Whole Life Cost or constant performance/benefit. Such an approach would enable a comparison on an equitable basis, thus leaving the difference in one dimension. A few large, costly, high-capacity systems at one extreme would be compared with many, small, low-capacity systems at the other.

In Figure 21.4 the quantity of systems has been adjusted until all the options have the same cost, then the effectiveness has been calculated. In the second scenario, the number of systems has been adjusted until the performance (in terms of the service delivery effectiveness) is constant; then the Whole Life Cost of each option has been calculated.

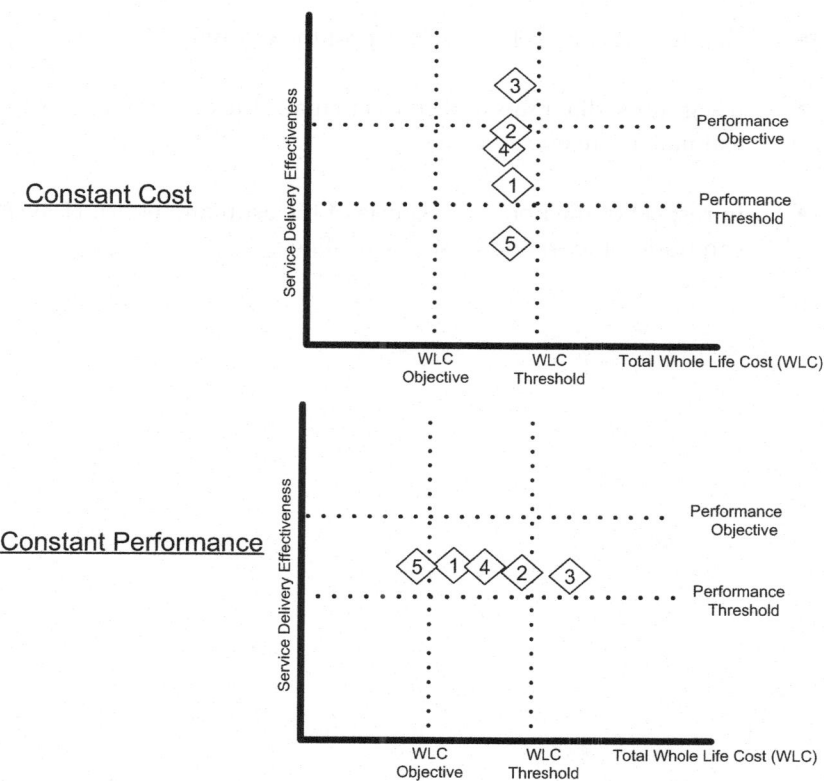

Figure 21.4   Constant cost or constant performance

Summary

In this chapter you have been exposed to the need to present cost estimates in a meaningful way:

- timing is important for any report produced after an estimate;

- consider and answer the question asked; don't answer the most convenient question for which data is at hand;

- ensure the estimate is put in context of its assumptions and exclusions;

- clearly and comprehensively report the estimate;

- use graphs, diagrams and charts to make the communication of the estimate more effective; and

- in respect of the non-cost aspects of the estimate report, IT systems can trade-off cost and business resilience.

22

How to ... Adopt Parametrics

Getting started on a parametric campaign is not as easy as it might seem. This chapter deals with techniques to tackle the top and bottom of the chain of command.

Process Re-engineering

Parametric model vendors have considerable expertise in all areas of Cost Engineering. The business of re-engineering the processes of an organization is time consuming and requires considerable dialogue to ensure that everyone is brought into the new processes.

Change is the single most important element of successful business management today. Change will be extremely important in achieving the aspirations of the Cost Engineering department initiative. To remain competitive, organizations (and individuals within them) need to adopt a positive attitude to change. Ignoring or trivializing a changing trend can be costly in terms of the budget secured for that organization.

Change management requires planning. Initially, two questions need to be asked: 'Where are we now?' and 'Where do we want to be?' The answers to these two questions determine the gap to be navigated. The next step is to agree the most effective way of changing the organization in order to navigate the gap and arrive at the answer to the second question. The secret is managing this process without losing the value that customers place on the current service that Cost Engineering provides.

Most organizations strive to exist by developing a stimulated, well motivated staff who will accept change (in moderation) as the challenge that it should provide. There is a balance to be found between staff who are bored through lack of intellectual investment and stimulation, and the other extreme in which some staff have been so demoralized by change they cannot cope with any more (see Table 22.1).[1]

Training can be used to involve and prepare people for change. The environment that training provides makes staff more accepting of change when they are taken out of their everyday environment and into a situation where they expect to experience new and novel ideas. It is possible in this environment to:

- invite suggestions from everybody;

- hold frequent and informal discussions on working practices;

- involve people with change planning and implementation;

- manage people's expectations.

Table 22.1 Dealing with reactions to change

Types of negativity	What to do about them
Rational Misunderstanding of details of plan; belief that change is unnecessary; disbelief in planned change's effectiveness; expectation of negative consequences.	Involve everybody in quality-improvement teams to demonstrate effectiveness of managed change. Project what would happen if the change program was not introduced. Explain plan with greater clarity and detail. Institute a bottom-up program for reorganizing systems and processes.
Personal Fear of job loss; anxiety about the future; resentment at implied criticism of performance; fear of interference from above.	Present a scenario showing the anticipated benefits of the main change. Stress much-improved job prospects for the future for everyone. Present plans for improvements that people are likely to find positive and exciting. Accept management responsibility for past failures.
Emotional Active and/or passive resistance to change in general; lack of involvement; apathy towards initiative; shock; mistrust of motives behind change.	Demonstrate that the new policy is not merely 'a flavour of the month'. Stage a series of meetings to communicate details of the change agenda. Show, with examples, why the old ways no longer work. Explain the reasons for the change, and promise involvement. Be completely honest, and answer all questions.

---

1 Taken from *Essential Manager's Manual*, R. Heller and T. Hindle, DK Publishing.

Model Acquisition

It is unthinkable that an organization would go out and acquire an engine, gearbox, chassis, hull, avionics, radio, radar, missile or any piece of equipment without establishing a detailed requirement (see Figure 22.1). But for some reason, many organizations procure a cost model, or develop their own cost model, in a very *ad hoc* way.

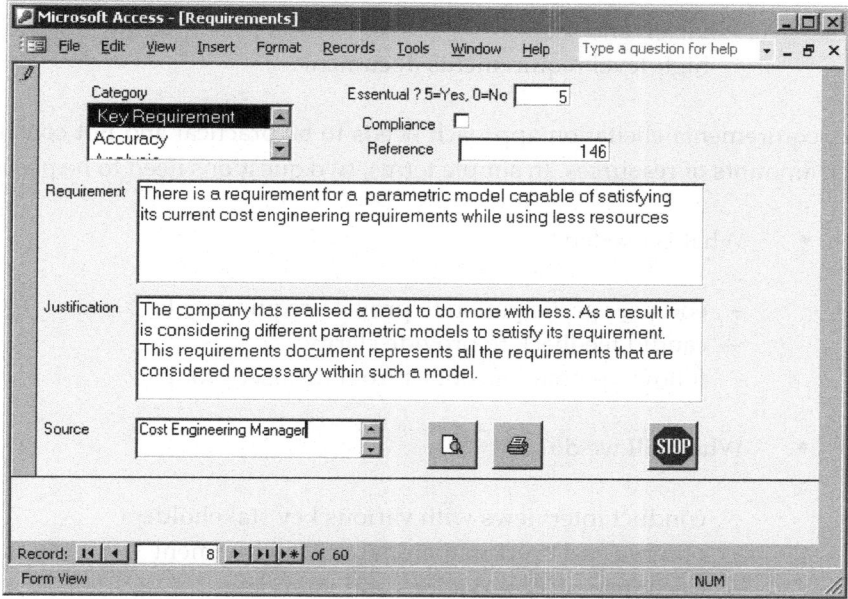

Figure 22.1   User requirements database

Model vendors would certainly advocate the need to conduct a requirements gathering exercise prior to the acquisition of a model. Moreover, having preformed requirements gathering exercises for other customers, they would be pleased to assist or advise anyone who is thinking of conducting the acquisition in a professional way.

The requirements elicitation objectives are best kept simple, they include:

- validating the organization's vision of the solution;

- eliciting and capturing:

  - 'as is' and 'to be' business processes of the organization
  - high-level end-user requirements
  - system requirements (non-functional)
  - current and desired tools for integration

- deliverables should include:

  - vision and scope
  - high-level requirements document

The requirements elicitation approach needs to be practical and not consume huge amounts of resources. In simple terms, two questions need to be posed:

- What is needed?

  - end user participation to provide requirements
  - candid feedback to provide clarity
  - follow up point of contact from the user groups

- What will we do?

  - conduct interviews with various key stakeholders
  - observe end users in their natural environment
  - document all observations and interviews
  - validate the requirements prior to study completion

Once the requirements have been determined, a sensible competition can be held. A Request for Proposal (RFP) should be issued and a proper evaluation to determine value for money conducted – remember, you can't get something for nothing. Is a third generation parametric model required or will a second generation parametric model be adequate? The case must be argued with senior management for a sensible budget for this exercise. The licence cost for a model and the cost of hardware, training and personnel to run it should be included.

Once the cost model has been selected, implementation of the real project can begin. The staff can be recruited and trained and their personal objectives set according to the implementation program.

Model Implementation

This section represents the cumulative experience of many cost estimating professionals and provides guidance regarding the most effective means of establishing a parametric cost estimating methodology.

This solution consists of two plans:

1. Parametric Implementation Plan.

2. Calibration and Analysis Plan.

Creating a Parametric Implementation Plan is crucial preparation for the long-term stability of the cost and schedule modelling that is to follow. The Parametric Implementation Plan has a strong focus on the organization of the work to be undertaken and the information flows needed to make parametric estimating a success in any business.

The Calibration and Analysis Plan is essential to ensure that the commercial parametric model is tailored to meet the specific accounting, organization and productivity requirements of the business. This is achieved in a series of short activities reviewing the technologies developed and produced by the business.

The combination of the two plans will enhance the capabilities and performances of the model and also improve the return on investment that can be expected.

The Project

No professional engineer would embark on the procurement of a ship, aircraft, satellite, vehicle, system or equipment without first initiating a project and producing a plan. The introduction of parametrics into an organization should be conducted with the same rigour and professionalism as any other procurement.

The requirement for the following needs to be considered:

- a project plan;

- a team;

- a requirements document.

PROJECT PLAN

Assuming the introduction of the parametric estimating methodology is one of your personal objectives, a project plan for the next 12 months is an excellent way to proceed.

Determine the objectives for the project over the next 12 months and plan the activities using the start-up plan below to determine what you need to accomplish. Once the plan is complete, you will be able to determine rationally the resources that will be required to execute the objectives. Resources should include staff time and licence costs.

TEAM

The introduction of a parametric estimating methodology cannot be completed in isolation. It will be necessary to call on the assistance of finance staff, program staff, management, business staff, estimating staff, and so on. It is important you recognize this from the outset and proceed with your project with the support and blessings of these other staff. If you embark on this project without the whole organization behind you, the introduction will be more difficult.

Use a kick-off or launch meeting at the beginning to *sell* the idea of parametric estimating to your colleagues. During the project, communication will be essential, remember to provide an update of the project's progress and brief staff at regular stages throughout the project.

A steering committee or management committee may be helpful to spread the load when decisions need to be made and to provide access to resources.

REQUIREMENTS DOCUMENT

Your organization must have reason to change. Ask yourselves: Why adopt parametric estimating? What is the requirement? The requirements document will ensure that the project does not lose its focus and purpose. It will provide the backbone of any resource discussions and justification. It will be an essential reference document in any business case.

Parametric Implementation Plan

The following is intended to be a guideline to assist trained users of parametric cost models with the introduction and implementation of this capability in their company. It provides hints regarding general points to be considered during the introduction. These guidelines should not be considered as a complete list of necessary actions as they will vary from company to company.

INTRODUCTION WITHIN THE ORGANIZATION

Make all staff involved in cost and schedule estimating (for example, projects, sales, program-management, engineering, production, and so on) aware that a tool is available to help them in their estimating activities. This can be achieved through meetings, presentations, memos, the company newsletter, printed documentation, and so on. Utilize a picture of yourself on the training course to make an impact.

PRODUCTS, TECHNOLOGIES AND PROCESSES

Take time to review the company Business Plan. What are you likely to be asked to estimate in the next two or three years? What are the new technologies in which the company is investing? Just because there are huge amounts of data on old products does not necessarily mean these are good candidates for estimating, especially if the company is changing strategy and moving out of that product area.

Consider:

- Designing a table to list the equipment produced by the company, with identification of old and new products. This will then form the basis of a knowledge database and prioritize effort.

- Spending time to identify the technologies such as digital electronics, composites, hydraulics, hyper-frequency, and so on with an indication of the operational specifications.

- Identifying the engineering and manufacturing processes such as CAD tools, NC machining, high-speed machining, casting, forging, and so on.

COST STRUCTURE

Determine the company cost structure. This will allow for correct comparisons between the parametric outputs and in-house estimates or actual costs. A correspondence table will need to be built to match the internal figures generated for activities to the parametric ones.

ECONOMICS

Adjustments can be made to the parametric escalation table to make the values consistent with your organization's desired inflation and escalation rates for future years. Research the most appropriate source of escalation data and be consistent, create an escalation table for your organization, involve the appropriate authorities in checking your interpretation of their data.

It is also possible to create a set of organization-specific tables for the organization's labour rate, overhead, overtime, G&A and fee rates.

AREAS NOT COVERED BY THE MODEL

At the beginning of the cost research process you need to define the boundary of the cost model that you are going to consider. Anything outside that boundary will be excluded and should be estimated by some other means. These additional estimated resources may be added to the parametric estimate as:

- *purchased goods* – unit costs which are integrated with other items;

- *purchased service* – labour hours which are integrated with other items;

- *other costs* – costs which are not integrated.

Civil engineering, consultancy, training, furniture, textiles, lighting, power or water distribution and fuel are all examples of items which might be estimated outside a parametric model and which will need to be estimated separately and added to the parametric model. Once the cost of these exclusions is known, you are able to include them in the model results.

HISTORICAL TECHNICAL DATA

You will need to acquire past program technical data for:

- hardware technical parameters such as weight, volume, number of parts, tolerances, hogout, and so on for existing equipments in order to develop guidelines useable on the new projects;

- software technical parameters such as Function Point count, Source Lines of Code count, Functional Complexity, and so forth for existing software programmes to assist in the estimation of new projects.

The purpose of this process is to be able to determine the input parameters for new projects which are not yet available. Setting up a parametric database could follow on from this data gathering and will be composed of the analysed items and their technical parameters.

COLLECTION OF HISTORICAL COST DATA

Historical cost data should be acquired for each type of product, technology, program or process. This cost data can be taken from historical or current projects, whether they come from cost control or analytic cost estimating. Calibration of a parametric tool is dependent on good quality historical data. This cost data will then be entered in the parametric database.

PRODUCT CALIBRATION

From the information above, together with your training, it will be possible to conduct product calibration. The output will be Organization Productivity for software and Manufacturing Complexity for hardware. Armed with this information you will be able to benchmark your organization against average industry.

Part of the product calibration process is the determining of the calibration factors for the parametric model calculators. These should be categorized by product-type, technology and process, and a record should be created for future reference. This product calibration should be performed in conjunction with the schedule calibration.

ORGANIZATIONAL CALIBRATION

To ensure that the parametric model outputs are aligned with your organization's accounting and organizational structures, it is possible to set the parametric model multipliers. Following the training, this capability will adjust the parametric results to reproduce the magnitude of the activity and resource that matches your expected estimates.

PARAMETRIC QUESTIONNAIRE

A quick and convenient method of seeking information for the parametric model is in the form of a questionnaire. This can be sent directly to the source of information or populated as part of a structured interview. Questionnaires can also be given to suppliers, subcontractors, engineering staff, and so on, for data gathering.

Adapt a tool vendor's questionnaire to your company with your logo or begin the questionnaire to suit your situation. It can be used equally well for historical data gathering, supplier's information or for future bids.

MODELLING TECHNIQUES

Parametric models are a collection of mathematical models. If you are consistent with the model the estimated cost and schedule results will be consistent; if you keep putting rubbish in you will get rubbish out. Use parametric tools in the calibration task the same way when you estimate.

Establish a set of rules; agree how you are going to apply parametrics and stick to it. For example, always use kilograms or pounds, always apply schedule effect or not, and so on.

APPLICATIONS

Consider the way in which you are going to apply parametrics. Parametric cost estimating can be applied to many complex problems. Asking your vendor support staff or tutor for an explanation of the following applications or solutions:

- quality assurance (parametric versus detailed);

- risk analysis;
- knowledge management;
- productivity tracking;
- competitor assessment;
- supply chain coherency;
- Historical Trend Analysis;
- collaborative working (RACI);
- influencing project strategy:
  - acquisition strategy (COTS/in-house)
  - support strategy (contractor support/in-house support)
- Systems of Systems (SOS);
- supplier assessment;
- Predictive Earned Value Management (PEVM);
- Cost as An Independent Variable (CAIV) or Design to Cost;
- estimating, forecasting, budgeting;
- technology insertion;
- Whole Life Cost (WLC) or Total Cost of Ownership (TCO);
- Business Resiliency;
- Through Life Estimating;
- affordability management;
- development of cost estimating models.

INFORMATION NETWORK

You will establish an information network covering sources of input to the parametric model: weight engineers, software engineers, accountants, schedule experts, and so on. Appreciate these contacts and foster them, you will need their cooperation during the estimating process to allow the quick population of input parameters, to solve problems or to answer questions. Provide feedback on your progress to them. Build a list of contacts containing their area of expertise, name, building, location, telephone number and email.

Calibration and Analysis

The calibration and analysis of a user's organization is a continuous process (as previously discussed in Chapter 8).

Education

TUTOR-BASED TRAINING

Training is an important element of the adoption process. Most courses are structured to cater for different learning methods: people who learn from doing and people who learn best from theory. Effective courses are a blend of discussion, teaching, presentation and practical exercises which ensure that students leave fully equipped with the skills to perform parametric estimates and, more important, the understanding of how these parametric estimates are produced. Understanding is the key to producing robust parametric estimates: it is only a convincing estimate when the Cost Engineer can communicate how it was derived. This process starts with the Cost Engineer having a full and convincing understanding of the tools, confident and ready to apply their newly acquired knowledge and skills.

As the traditional end-of-course speech goes: 'This is the end of the course, but the beginning of a relationship.' Parametric vendors will provide support on an ongoing basis to customers around the world. Refresher training for life is freely available to students who have a fresh need for parametric skills after perhaps a period of alternative engagements.

The best type of parametric training is modular, enabling students to learn how to create a project, change labour rates, alter escalation, apply risk analysis and so forth. These skills are learnt once and apply to all the cost models used.

WEB-BASED TRAINING

Web-based training augments tutor-based training, but is not a real substitute for face-to-face classroom style education. Web-based training does have its place though, in maintaining skilled cost professionals through the life of a program. Parametric tool vendors usually offer web-based training free of charge to licensed users, to provide a revision capability to students. Web-based training has the advantage for some students of self-paced learning and practice.

Another advantage of web-based training is that it can be revisited at any time and from any place: while travelling, at an airport, or anywhere a web connection is possible. It is also a convenient way to ensure that you are updated and keep pace with product improvements and tool enhancements.

HELP DESK

As well as a comprehensive package of training, help desk support has always been available to Cost Engineers. This technical support service is provided as part of the standard contract with quality parametric models. This is normally maintained and enhanced with a website containing technical White Papers and 'webinars' to ensure the customer has multiple means to answer any query they may have.

Client/Server Capability

Adoption of parametrics can be accelerated by the use of client/server installation. From Figure 22.2 it is possible to determine a structured approach to cost estimating. It is also possible to determine who in the organization is responsible for the activities that need to be executed prior to the estimate being completed.

	Senior Management	Task Manager	Mass Dept.	Design Dept.	ESTIMATOR	Finance	Customer	Commercial	Marketing
Activities or Decisions	I	R	A	A		R	C		
		I	R	R	A	R	R		
		I	C	R	A		C	R	C
		I	R	R	A		C		
	I	R	A	A	R	R		I	

R = Responsible	Many people	Work, reports or data required
A = Accountable	One person	'The buck stops here!'
C = Consulted	Two-way communication	Positive input, ideas or thoughts
I = Informed	One-way communication	Supplied with output

Figure 22.2 RACI matrix

Figure 22.2 provides a means of collaborative working to ensure a successful conclusion of an estimating task using manual documentation. Down the left-hand side of the RACI matrix are the activities to be conducted by an organization as part of a cost estimate. The following might be included:

- kick off meeting;

- prepare initial documentation, that is, Basis of Estimate document:
  - agree assumptions and exclusions
  - agree methodology
  - agree tool set, and so on.

- carry out estimates as agreed with customer;

- review all deliverables with customer;

- deliver what was specified in the original customer request.

Across the top is a list of the necessary functions which are, in some way, involved in the process at some stage. The body of the matrix sets out the RACI, the definitions of which are very straightforward:

- R = Responsible – identify people doing the work.

- A = Accountable – the buck stops here.

- C = Communicated – two-way interaction.

- I = Informed – one-way interaction.

A client/server implementation of a parametric cost model has many advantages to accelerate the use of parametrics – among them would be:

- share estimating models across teams or functions;

- encouraging all Cost Engineers to share a common Graphical User Interface (GUI) – saving time and cost on training;

- motivating subject matter experts (technical/program/financial) to contribute their Cost Drivers and parameter values directly and to populate the estimate;

- consistent cost modelling across the organization introduced by utilizing common cost models – saving time and cost on model maintenance – either:
    - commercial or
    - in-house or organic (custom-built models)

Figure 22.3 shows a client/server configuration for third generation parametric cost models. Multiple servers can be accessed by multiple users. This ensures that internal estimates are exploited quickly and efficiently. The best, most accurate cost estimates are based on an organization's past experiences captured in past estimates. Providing direct access to the cost model inputs to the appropriate expert ensures that confidence is increased in the basis of that estimate. Time is saved in recovering estimates, particularly during employee absence due to sickness, vacation or termination. Moreover, there is no confusion regarding the latest estimate produced. Furthermore, for convenience the parametric model can, and should, have the ability to 'briefcase' an estimate thus making it available locally on a laptop for off-site negotiations or shared storage on a server for communal usage.

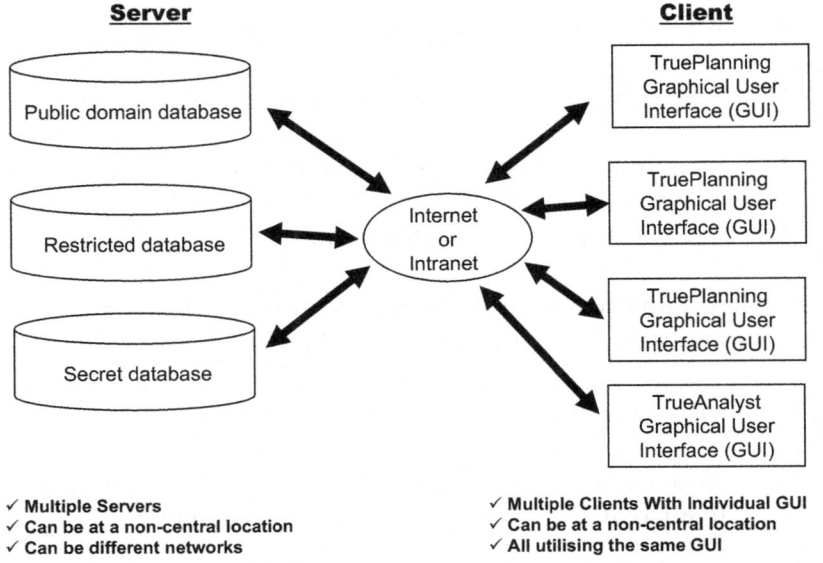

Figure 22.3   Example client/server configuration

Client/server configuration can provide a consistent, logical organization for rapid searches and analysis. It needs to be built on easily maintainable industry standard technologies (that is, SQL Database, Internet Browser, XML).

WORKFLOW MANAGEMENT

Once a configuration has been achieved that involves the cost estimates being stored on a central server, it is simple to envisage a system which interrogates that server and manages the projects stored. A web-capable system using Internet Explorer for administration ease of use will provide a great deal of capability and detail, including:

- creating and deleting system users;

- creating and deleting passwords;

- monitoring workflow;

- Cost Engineer responsible workload;

- percentage of the work completed;

- estimated value of projects;

- cost model(s) utilized;

- when the estimate was last modified.

If the workflow was able to monitor against the RACI matrix described earlier, it would bring together two capable tools to become a system which could manage workflow activities. It would define the stages of the activities to be performed and become an enforcer to aid quality assurance in terms of an *aide-mémoire*. Providing the ability to 'check off' the stages of work completed could help to balance workloads across the cost estimating function. Moreover, recording when and by whom the activities were completed has the ability to generate an audit trail.

The benefits of a client/server installed parametric cost model are numerous, but include:

- sharing of estimates;

- workflow monitoring;

- implementation of a RACI methodology but in a real time and efficient manner;

- managing the workload of Cost Engineering in an organization.

Summary

Changing an organization to accept parametric estimating can be challenging, particularly if it has not seen the wider benefit of Program Affordability Management. Treating its adoption as a program like any other program will help to focus activities on the smooth transition to include parametrics throughout the organization.

23

The History of Parametrics

Writing a history is a bit like producing an estimate, it is hard to do without putting a boundary around the subject. Like an estimate we need to know what is going to be included in the history and what is to be excluded or not going to be covered. For example, the Bible has the most commonly quoted and earliest references to estimating, 'For which of you, intending to build a tower, does not first sit down and estimate the cost, to see whether he has enough to complete it?' which appears in the New Testament (Luke 14:28). So estimating has been acknowledged as a worthy pursuit by some of the earliest stone masons and craftsmen. In this chapter I intend to focus on more recent history.

Estimating also covers different methodologies: detailed, analogous and parametric. In this chapter I will focus on parametric history. Even within this subject it is possible to find various types of parametric estimating models. These can be divided into organic, in-house or home-grown models created within an organization for the sole use of that organization, and commercial models built by a third party for others to utilize. For this chapter I intend the centre of attention to be the history of commercial parametric models.

The *NASA Parametric Cost Estimating Handbook*[1] states that: 'The origins of parametric cost estimating date back to World War II. The war caused a demand for military aircraft in numbers and models that far exceeded anything the aircraft industry had manufactured before. While there had been some rudimentary work from time to time to develop parametric techniques for predicting cost, there was no widespread use of any cost estimating technique beyond a laborious buildup of labour-hours and materials. A type of statistical estimating had been suggested in 1936 by T.P. Wright in the *Journal of Aeronautical Science*. Wright provided equations which could be used to predict the cost of airplanes over long production runs, a theory which came to be called the learning curve. By the time the demand for airplanes had exploded in the early

---

1 http://cost.jsc.nasa.gov/pcehhtml/pceh07.htm.

years of World War II, industrial engineers were using Wright's learning curve to predict the unit cost of airplanes.

In the late 1940s, the Department of Defense (DoD) and, especially, the United States Air Force began a study of multiple scenarios concerning how the country should proceed into the age of jet aircraft, missiles and rockets. The Military saw a need for a stable, highly skilled cadre of analysts to help with the evaluation of such alternatives. Around 1950, the military established the Rand Corporation in Santa Monica, California, as a civil "think-tank" for independent analysis. Over the years, Rand's work represents some of the earliest and most systematic studies of cost estimating in the airplane industry.'

In the 1950s Rand researched development and production Cost Estimating Relationships (CERs) for different types of aircraft and matured these techniques in the aerospace domain through the 1960s.

Until 1994 parametric cost estimating was chiefly employed for rough-order-of-magnitude estimating, cross-checking primary estimates and cost analysis. The Parametric Estimating Initiative (PEI) in April that year was the joint US industry and government (Defense Contracts Management Command (DCMC) and Defense Contracts Audit Agency (DCAA)) initiative to address the barriers to parametric estimating as a sole basis of estimate for proposal pricing. This milestone cleared the way to apply parametric estimating for primary estimates. Today, the number of companies that are bringing speed and accuracy to their processes with parametric estimating is growing.

Generations of Commercial Parametric Models

Commercial parametric models began with first generation models adopting the earliest mainframe computers. Prior to the adoption of computers, using graph paper, a pencil and a slide rule it was possible to derive parametric models, but the distribution of that knowledge for commercial use was difficult unless published in a book. With the adoption of computers in estimating it was possible to make a modest material gain from the licensing of commercial parametric models. This support allowed further research and more importantly effortless sharing of parametric solutions across the aerospace and defense community. The life time of these early parametric models spanned from 1975 to 1990.

Early Parametricians needed to be taught text editor and keyboard skills as part of the commercial parametric courses. Cost Engineers of the 1970s were still marvelling at the electronic calculator when the cream of the youngest, most talented Cost Engineers had their productivity slowed to a grinding halt by one finger and a keyboard. Estimates were prepared and processed overnight on the mainframe, only to find in the morning that a vital input was missing and the whole estimate would need to be run again the next night. However, intellectual property in the form of the trade secret algorithms contained within commercial models were deemed sufficiently valuable that management and Cost Engineers were prepared to accept mainframe computers as an imperfect solution.

Second generation parametric models began to arrive when personal computers were embraced in industry in 1988 and are still used today. As personal computers became powerful enough to deal with the thousands of mathematical calculations required in the parametric models, the software of the first generation models were ported onto these computer platforms. The second generation models also had the advantage of Graphical User Interfaces (GUI) and easier dialogue with the user. This led to the more efficient use of commercial parametric models and the further expansion of their application. The commercial parametric model became portable and could be taken to subcontractor or customer sites for negotiations.

Whereas the processing of the first generation parametric models was done centrally, using dumb terminals via 1,200 bits per second or 2.4 kbps modems to retrieve the results, the second generation commercial parametric models had to deal with the issue of protecting the trade secret equations within the programming. The software was compiled to run faster on personal computers, which led to the trade secret equations within the programming becoming scrambled – to deal with the issue of protection – which in term led to the problem of validation and verification. The term 'black-box' was commonly used by staff who found commercial parametric models unfathomable. In these second generation models the users were ultimately unable to see the mathematics or parametric algorithms; they had to depend on the quality assurance of the commercial parametric models vender.

Finally, at the turn of the millennium, third generation parametric models arrived on the market. Taking advantage of the latest client/server computer architectures combined with open parametric estimating frameworks, third generation parametric models enable commercial parametric models vendors to port their models into this framework and implement bespoke organization specific models. Within the open parametric framework these models (from different origins) can use the common cost engineering features of the framework like labour rates, escalation, risk analysis, reporting and so forth.

Cost savings on third generation models are possible because the same user interface is used for all the models whether software, IT or hardware. Training can be modular – with framework training conducted only once followed by model-specific training – thus, providing a more pleasurable user experience. Furthermore, validation and verification of the models is simpler as the open environment can be scrutinized by any authority to review the implementation of the cost estimating relationships.

Founding Father of Parametrics

Mr Frank Freiman, the recognized founder of parametrics, had been working on a series of equations or algorithms for estimating hardware equipment since the middle of the 1960s. This gave the Radio Corporation of America (RCA) an advantage over its competitors when comparing the timeliness and quality of its estimates.

Recognition of this advantage encouraged the US Department of Defense (DoD) in America to license the software which was used to derive estimating costs parametrically. This enabled the DoD to apply a little bit of science to a process that had hitherto been time consuming.

Figure 23.1   Frank Freiman – founder of parametrics

This was in 1975 when PRICE Systems became a division of the American company RCA for the purpose of licensing parametric cost models. It started a revolution in the areas of schedule and cost estimating and schedule and cost control. The 'Freiman Curve' (see Figure 23.2) was the value proposition that helped to launch the new company. Freiman recognized that parametrics could support good-quality estimates which were the key to influencing the cost outcome of a project. A cost under-estimate would lead to an increase in the final cost out-turn due to a lack of resources, irrational quick decisions and unrealistic expectations. However, over-estimating the initial cost estimate had the same consequences due to under-utilization of staff, acquiring excess capacity and non-competitive pricing which led to higher cost out-turn than expected. Parametrics provided checks and balances for the initial cost estimate and improved its quality.

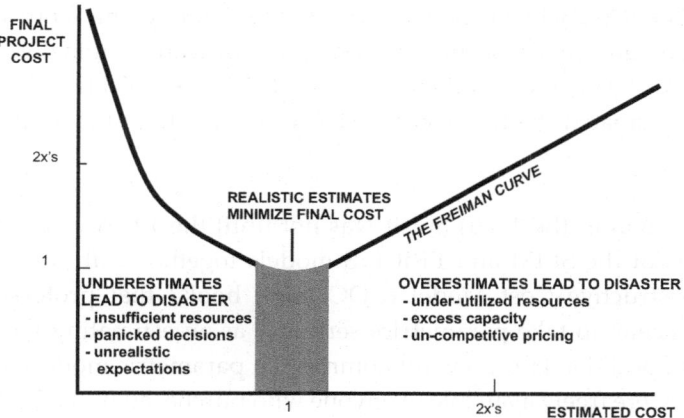

Figure 23.2   The Freiman Curve

PRICE Systems business looked around the world and found non-US companies it could help. As a result, it became a global organization with offices throughout the United States, Europe and the rest of the world spreading the parametric technique.

With the addition of non-hardware models, demand quickly grew together with the maturity of the processes, from a methodology which showed promise through to a technique which became the sole Basis of Estimating (BoE) in some organizations. A PRICE Systems User Group soon became established, with people from different organizations coming together to network and compare estimating techniques. At the same time other parametric models began to appear.

In 1980, recognizing the maturity of the industry and the need to represent other parametric cost models, the International Society of Parametric Analysts (ISPA) split from the PRICE User Group (PUG).

Commercial Parametric Models

Commercial parametric models naturally split into those capable of estimating software programs and those capable of estimating non-software programs. Within commercial software estimating models there have been a number of systems developed from diverse independent research organizations (see Figure 23.3). Lawrence H. Putnam, Sr[2] developed the work of the IBM Corporation which had been produced by Peter V. Norden[3]. This would later become the Software Lifecycle Model (SLIM) by Quantitative Software Management (QSM) Incorporated in 1979.[4] Separately, the Doty Associates model[5] made an impact on the estimating of software, originally developed for the US Air Force. Independently within PRICE Systems Frank Freiman and Robert Park[6] created the first commercial software estimating model PRICE S in 1977.

These began in the 1970s but it was not until the 1980s when there were refinements of the SLIM and PRICE S models together with the publication of the Constructive Cost Model (COCOMO) book,[7] that professional Cost Engineers began to take parametrics seriously as an estimating methodology for software and demand grew for commercial parametric models. COCOMO algorithms were freely available to anyone who purchased Barry Boehm's book and this created a small industry in itself that began coding the COCOMO algorithms into GECOMO,[8] COSTAR,[9] REVIC[10] and others.

---

2   L.H. Putman (1978), 'A General Empirical Solution to the Macro Software Sizing and Estimating Problems', IEEE Transactions on software engineering.
3   P.V. Norden (1970), 'Useful tools for Project Management', in *Management of Production*, Penguin Books
4   http://www.qsma.com/about_history.html.
5   J.R. Herd, J.N. Postak, W.E. Russell, and K.R. Stewart (1977), 'Software Cost Estimation Study – Final Technical Report', Vol. I. RADC-TR-77-220. Rockville, MD: Doty Associates, Inc.
6   F.R. Freiman and R.D. Parks (1979), 'PRICE Software Model Version 3: An Overview', in Proceedings of the IEEE-PINY workshop on Quantitative software models, IEEE Press.
7   B.W. Boehm (1981), *Software Engineering Economics* (Englewood Cliffs, NJ: Prentice Hall).
8   From GEC-Marconi.
9   www.SoftstarSystems.com.
10 Revised Intermediate COCOMO (REVIC).

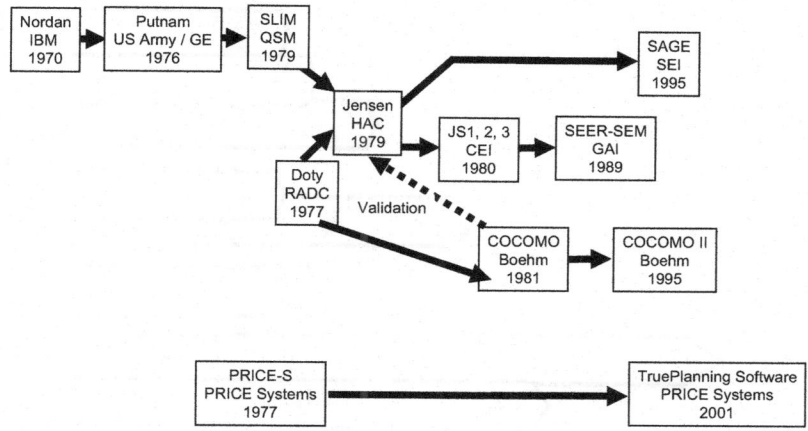

Figure 23.3   Commercial software parametric models
Source: Adapted from a chart completed by Dr Randy Jensen, 2006 ISPA Conference, 'Have We Learned Anything in the Last 20 Years?'

In the 1990s the SAGE model[11] was implemented by Software Engineering Inc. which added to the Jensen model the influences of project management and personnel characteristics. Software Evaluation and Estimation of Resources – Software Estimating Model (SEER-SEM)[12] was introduced by Galorath Associates as a development of the Jensen software model.

The first third generation parametric models was TruePlanning for Software introduced in 2003 a derivative of the PRICE S model with enhancements due to the client server, activity-based modelling approach and comprehensive Commercial Off The Shelf (COTS) algorithms.

On the hardware front, the parametric estimating methodology developed commercially through PRICE Systems on the east coast of the United States (see Figure 23.4). Noteworthy, PRICE Systems innovators include William E. Rapp for the Hardware Lifecycle model (PRICE HL), Anthony A. DeMarco for the Micro-electronics model (PRICE M), John Long for the Total Ownership Cost (TOC) model and Arlene Minkiewicz for the TruePlanning for Hardware model.

---

11   Software Engineering Inc (1995), *Sage User's Manual* (Brigham City, UT: Software Engineering, Inc.).
12   Galorath, Inc. (2001), *SEER-SEM Users Manual* (El Segundo, CA: Galorath Inc).

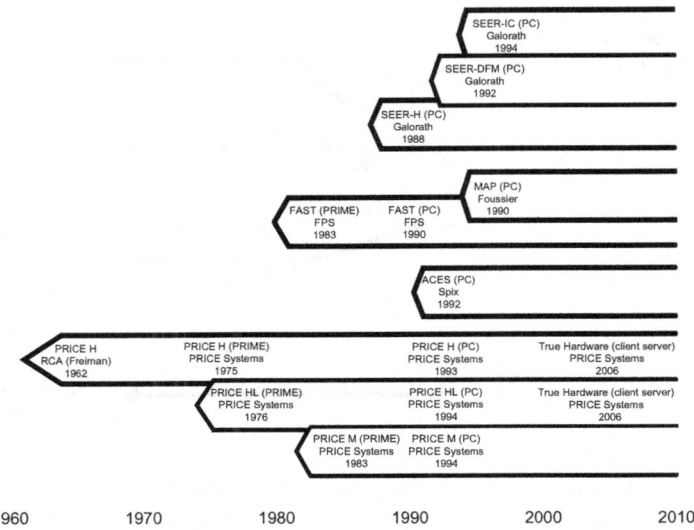

Figure 23.4 Commercial hardware parametric models

Frank Freiman left PRICE Systems and launched Freiman Analysis of Systems Technique (FAST) in 1983. Over several years, other staff have left PRICE Systems to spread the parametric methodology. Prompted by the lack of implementation of the models on a personal computer, Herbert Spix implemented the ACES cost estimating model in 1992.

On the west coast in 1979 Dan Galorath started a consulting business (Galorath Incorporated) and introduced the SEER suite of models in about 1988.[13] Peter Korda joined Galorath Associates from PRICE Systems in 1992 to develop the Design for Manufacture (SEER-DFM).

The Future?

What about the future? What directions will parametric estimating take over the next few years? Advances in information technology and the Internet will have a dramatic impact on parametric estimating. The ability to share knowledge and historical information, cost, schedule and technical matters will influence

---

13 See 'SEER for Software from Galorath: Setting the Standard for 20 Years', <http://www.golorath.com/index.php/news/press/seer-for-software-from-galorath-setting-the-standard-for-20-years>.

the techniques of the future. More statistics and metrics will be captured – but will be even more accessible. Training will be more interactive and web-based.

The parametric cost modelling will become more complex – but easier to use. Third generation models will result in the proliferation of models available and shared in the community, but with this comes responsibility. There will be more focus on cost models in terms of validation and verification of cost models used. It will no longer be acceptable to utilize ad-hoc spreadsheets. Peer review and scrutiny of the models will provide a greater level of confidence in these models.

There will be more complex decisions to make – but each one will be supported by parametric measurement, analysis, and forecasting. As more programs are required from the same or diminishing budgets, the level of accuracy required will be important to ensure all stakeholders can be accommodated. Parametric tools will become integrated into the tool capabilities of other engineering disciplines. Parametrics will be applied not just to systems and parts, but also to personal productivity. Estimates will not be the sole responsibility of individuals, but will be the collaborative efforts of integrated teams comprised of diverse and geographically dispersed experts.

The future of parametric cost and schedule estimating will be fun.

Index

A
ACES 298
accounting structures 80
accuracy 5, 31, 37, 51, 71, 85, 90–1, 238, 245
Acquisition strategies 159, 212
Activity Based 20, 63, 121, 224, 297
Actor 136
Actual Cost of Work Performed (ACWP) 60
Adapted code 128
advantages 59, 142, 146, 159, 222, 244, 287
Affordabilty management *see* Program Affordability Management
aircraft carrier 230, 239, 261
Albrecht, Allan J. 130
Alternatives *see* Cost as An Independent Variable (CAIV)
analogous estimating 17, 85, 106, 222, 263, 267, 291
Analytical estimating 31, 33, 85, 87, 98, 107, 222, 241, 267, 291
Anecdotal evidence 35
Ariane 142
automobile 99
Auto-generated code 128
Average Depth of Class in Hierarchy Tree (DIT) 134
average industry 15, 42–3, 47, 73, 281

Average Number of Children per Class 135
Average Tolerance 71

B
banking 268
basis of estimate (BOE) 95, 222, 253, 286, 292
batches 170, 179
benchmarking 35, 36, 44, 262
Benefit Assessment 267
bespoke 21, 293
Bid 2, 7, 31, 35, 87, 91, 195, 200, 253
bid/no-bid decision 31
black box 18, 293
Boehm, Barry 296
bottom-up *see* analytical estimating
broken production 170
Budget 31, 65, 73, 82, 93, 141, 170, 195, 200, 246, 276
Budgetting *see* Budget
Budgeted Cost at Completion (BAC) 60
Budgeted Cost of Work Performed (BCWP) 59
build-to-print 207
Business Plan 41, 89, 279
Business Resilience index 267

C
calculators 25, 26, 43, 61, 79, 83, 115, 123, 137, 140, 201, 281

calibration xvii, 5, 13, 15, 43, 51, 71–4, 79–80, 260, 277, 281-2, 284
Calibration factor 79, 281
Capability Maturity Model (CMM) 26
capital facilities 113
Cause and effect 63, 143
Cost Analysis Requirement Description (CARD) 150, 196, 265
Change management 273
Chief Scientist 7, 256, 260
civil infrastructure 118
Client/Server 20, 97, 108, 239, 285, 287–8, 293
COCOMO II 85, 99, 296
coefficient of variation 76
collaborative environment 106, 109
Commercial Off the Shelf (COTS) xvii, 25, 127, 138, 159, 212, 218, 297
communication issues 216
Compensated Gross Tonnage (CGT) 36
Competition 42, 44, 138, 167, 276
Competitor assessment/analysis 49
competitor's websites 46
complexity calculators see calculators
compounding errors 72, 90
Computer Aided Design (CAD) 89, 102, 202, 236, 259, 279
Concept of Analysis (COA) 195
confidence report 153
Consumer Price Index (CPI) 228, 255
Contractor Logistic Support (CLS) 174
core equation 6, 11
corporate knowledge 1, 110, 253
COSTAR 296
Cost account 65
Cost Accounting Standards 96

Cost analyst see cost engineer
Cost as An Independent Varaiable (CAIV) 7, 76, 102, 197, 283
Cost Control 2, 57, 281, 295
Cost Data Variability Analysis 150–1, 156
cost density xvii, 12-3, 16, 36, 43, 49, 73, 93, 190
Cost Drivers 5, 9, 28, 52, 53, 113, 160, 163, 189, 225, 239, 267, 287
Cost Engineer 3, 5, 17, 24, 46, 71, 100, 106, 195, 253, 284
Cost Estimating Relationships (CER) 5, 32, 89, 223, 225, 229, 231, 254, 294
Cost estimator see cost engineer
Cost forecaster see cost engineer
cost improvement curve 172, 222, 228, 234
cost influences 28, 184
Cost research 6, 15, 73, 85, 166, 221, 226, 238, 256, 280
Cost versus benefit 270
COSYSMO 99
COTS components see Commercial Off the Shelf (COTS)
customer furnished 11, 28

D

Data Centre 268
data dictionary 243, 249
data gathering 74, 227, 241, 244, 250, 255, 259, 282
Data manager 256
Data Variability Analysis 151
DEF FORM 143 37
Definition 53, 58, 105, 129, 147
Deleted code 128
DeMarco, Anthony A. 297
dependent variable 6
Deployment and Employment 251

derivation of the estimate 267
Design Authority 207
Design to cost (DTC) 197,
    202, 283 see CAIV
Detailed estimating see
    Analytical estimating

E
Earned Value Management
    (EVM) 1, 34, 57, 63
Ease of use 8, 240, 288
economics 13, 93, 228, 280
Economic analysis 196
efficient management 32
Engineering Changes
    Notices (ECN) 34
Engineering complexity 34, 114, 124
Estimate At Completion
    (EAC) 61, 66
Estimate To Completion (ETC) 66
estimating framework 107, 121
estimating process 23, 87, 263, 284
estimating size by Analogy 139
estimating software 99, 127, 296
Estimating Through Life 105
Excel see MS-Excel

F
FAST 297
Federal Acquisition Regulation
    (FAR) 96
Field test 242
fighter aircraft 93, 208
Financial Analysis 196
fly-by wire 124
forecasting 92,
Freiman Curve 295
Freiman, Frank 294
FRisk methodology 152
Function Points (FP) 25,
    130, 162, 215

Functional Size 25
Furnished Items 11, 28, 34

G
Gas Turbine 200
Galorath, Dan 298
Gaffney, John E. 130
GECOMO 296
Government Furnished items (GFX)
    34 see also Furnished Items
GPS Receiver 9, 248
Graphical User Interface (GUI)
    3, 24, 27, 223, 287
grass-roots see analytical estimating
Gross estimate 107
Guided Missile see missile

H
Help desk 285
Historical Trend Analysis (HTA)
    92, 94, 119–20, 190, 283
HL Questionnaire see Questionnaire
home-grown 28, 139, 221, 291
History 291
hybrid module 13
hypothesis 11, 200, 231, 237

I
implementation 201, 240,
    254, 277, 279, 287
impact 6, 18, 101, 141, 144,
    147–8, 198, 233
In Service Date (ISD) 93, 176,
    183, 200, 232, 254, 261
Independent Baseline
    Review (IBR) 69
independent variables 6
Information Technology (IT) 19,
    99, 108, 153, 208, 268
Initial Operating Capability
    (IOC) see in service date

insurance 268
Integrated Design to cost (IDTC) 204 see also CAIV
Integrated Project Teams (IPT) 26, 92, 173, 247
Integration complexity 217
International Function Point User Group (IFPUG) 26, 130
International Society of Parametric Analysts (ISPA) 226, 296
interview 146, 241, 245, 276, 282
Investment Appraisal (IA) 196

J
Jacobson, Ivar 136
Joint Strike Fighter (JSF) 13

K
Key performance parameters 217
Kick-off meeting 31, 278, 286
KnowledgeManager 230
Knowledge Management 2, 54, 108, 253, 283
Korda, Peter 298

L
labour rates 107, 165, 174, 285, 294
Launch meeting see kick-off meeting
lead system integrator (LSI) 211, 213
learner curve see cost imporvement curve
Life Cycle Cost (LCC) 173, 197, 199, 241
Long, John 297
lots 170 see also batches

M
Mainframe computer 292
major proposals 52
Maintenance Concept 175, 251

Manufacturing complexity xvii, 12–3, 16, 36, 43, 44, 49, 74, 78, 118, 190, 199, 261, 281
Market forces 18, 119, 166
Master Data and Assumptions List (MDAL) 150, 196, 265
Mathematical model 221, 282
Mean Time Between Failures (MTBF) 33, 173, 204, 244, 249
Mean Time To Repair (MTTR) 174, 191, 244
Mil. Std. 881 258
Military aircraft radio 176
Minkiewicz, Arlene 297
Missile 47
Mitigation Actions 148
Mobile Telephone 23
modified code 128, 140 see also adapted code
Modular Architecture 188 see also modularity
Modularity 171, 184, 189
Monte Carlo Analysis 82, 151
most likely 147, 155
motor control card 17
MS-Excel 18, 100, 123, 247
Multi-colinearity 230
Multinational 165
multiple methods 85, 103
multiple regression 229
munitions 124

N
NAFCOM 15
NASA 15, 111, 291
National Audit Office (NAO) 92
New code 128
non-optimum schedule 167
normalization 13, 36 49, 51, 73, 81, 227
normalized cost 93 see also Manufacturing Complexity

normalized cost density *see also* cost density
Number of Top Level Classes (TLC) 133

O
obsolescence 182
openness 184
Operating and Support (O&S) 36, 106, 174, 198, 221, 237, 241
Operating specification 124
Operational Plan 41
Operational Analysis (OA) 260, 267
operational scenario 216
optimistic 143, 153, 268
optimum schedule 39, 167, 173
organic *see* home-grown
organic sizing 139
Organization Breakdown Structure (OBS) xvii, 52, 58
Organization for Economic and Commercial Development (OECD) 36, 228
Organizational calibration 80, 282
Organizational Productivity xvii, 26, 42, 45, 65, 74, 127
Output 3, 5, 9, 27, 71, 78, 91, 125, 154, 191, 221, 225, 239, 251, 267, 280
Overall Tolerance 71, 90

P
paper studies 122
Park, Robert 296
parametric 3, 5, 7, 10, 16, 23, 31, 42, 52, 61, 63, 85, 107, 221, 273, 291
Parametric consultants 2
Parametric Cost Estimating Initiative (PCEI) *see* Parametric Estimating Initiative (PEI)
Parametric Cost Estimating Handbook 291

Parametric Estimating Initiative (PEI) 9, 95, 292
parametric questionnaire *see* questionnaire
Parametrician *see* cost engineer
pessimistic 153, 155
Planning 141, 145, 150, 273
POPs *see* Predictive Object Points (POPs)
Predictive earned value management (PEVM) 63
Predictive Object Points (POPs) 25, 132
preferred suppliers 37, 45, 52
presentation 260, 265, 270, 279, 284
PRICE Systems xv, 20, 38, 98, 247, 295
PRICE Systems User Group (PUG) 295
prime contractor 91, 207
Printed Circuit Board (PCB) 43
Proactive estimating 89
Probability 82, 141, 147, 202
Probability Impact Grid (PIG) 148
process 2, 23, 33, 44, 74, 87, 145, 152, 196, 199, 204, 207, 226, 250, 273, 279
Process re-engineering 273
Processes *see* process
procurement function 53
procurement strategy 28, 172
Product Breakdown Structure (PBS) xviii, 9, 23, 46, 53, 91, 99, 113, 121, 177, 186, 196, 249, 257
Product calibration 51, 73, 281
productivity 7, 12, 17, 36, 42, 44, 49, 63, 73, 79, 93, 108, 165, 190, 228, 299
productivity index 44
Productivity tracking 7, 42, 44
program xviii, 1, 34, 57, 62, 82, 105, 111, 120, 127, 144, 155, 166, 169, 177, 181, 254

Program Affordability Management
    (PAM) 1, 195, 289
programme xviii, 7, 113, 129, 281
Project Plan 41, 160, 277
Project strategy 159
proposals 32, 36, 49, 94
Prototype 113, 116, 122, 169, 236
purchased items 29, 37, 51
    see also COTS
Purchasing Power Parity
    (PPP) 165, 228, 234
Putnam, Lawrence H. 296

Q
qualitative benefits 51
Quality Assurance 85, 106,
    238, 289, 293
Questionnaire 37, 74, 246, 282
questions 10, 53, 79, 178, 194, 255, 273, 276

R
RACI 286
RAND 223, 292
Rapp, William E. 297
Re-engineering 273
reference range 49, 76
Reinvention Laboratory 9, 95
Reporting 8, 91, 150, 294 see also output
Requirements Database 275
Request For Proposal (RFP) 77, 195, 276
Residual 13, 74, 232
Responsibility 54, 58, 143, 208, 286
results 15, 23, 27, 67, 71, 75, 80, 126,
    150, 156, 202, 260, 265, 270
return on investment 5, 49, 73, 277
Reused Code 128
REVIC 296
Risk 1, 82, 102, 111, 141, 164,
    196, 201, 208, 262, 268
Risk Analysis 7, 82, 107,
    141, 150, 203, 262

Risk and Uncertainty 140
Risk Methodology 150
Risk Mitigation 149
Risk Identification 146, 197
Risk Process 142, 145
Risk Prompt List 146

S
SAGE 297
satellite 13, 61, 141, 225, 277
schedule 3, 5, 10, 23, 27, 32,
    61, 73, 105, 126, 144, 167,
    173, 221, 225, 235
Schedule effect 35, 168, 282
Schedule Estimating Relationships
    (SER) 32, 173
Schedule penalty see schedule effect
second generation parametric
    model 293
SEER 297
sensitivity analysis 8, 267
Should Cost 14, 17, 35, 77, 159
simulation 82, 100, 213, 242
size of software 25, 127
Skilled People 1
SLIM 296
SLOC see Source Lines of Code (SLOC)
SMART Procurement 92
software 23, 85, 113, 125, 127,
    159, 186, 212, 281, 296
Source Lines of Code (SLOC)
    25, 66, 129, 215, 281
specifications 46
speed 5, 31
Spix, Herbert 298
Spreadsheets 5, 18, 27, 74,
    106, 145, 299
stakeholders 92, 98, 211, 217, 276, 299
Statement Of Work (SOW) 58, 196
Strategy 7, 28, 161, 204, 207, 211
Supplier Assessment 2, 7, 36, 49, 51

INDEX

supply chain 52, 97, 159
System of Systems (SOS) 207
Systems Engineer 24, 46, 100, 159, 214

T
tables 13, 25, 118, 234, 280
target 31, 45, 50, 76, 198, 228
technical documentation 46
technical immaturity 218
technology improvement 51, 166, 231
Technology Insertion 171, 177, 181
Technology Maturity 13, 111, 167
Technology Maturity Date 119, 166
Technology Readiness Level (TRL) 111
Theoretical First Piece (T1) 228
Three point estimate 152, 155
Through Life Estimating 105, 223
Through Life Management Plan (TLMP) 173
Translated code 128
TrueAnalyst 20, 237, 288
Tolerance *see* accuracy
top down 23, 86, 98, 108, 224
Total Cost of Ownership (TCO) *see* Whole Life Cost (WLC)
Total Ownership Cost (TOC) *see* Whole Life Cost (WLC)
trade stand 46
Training 2, 96, 108, 274, 284, 287, 294, 299
TrueMethods 2
TruePlanning 2, 99, 107, 288, 297,
Truth in Negotiations Act 96
Tutor-based training 108, 285

U
unadjusted Function Point count 131
uncertainty 8, 82, 141
Unified Modelling Language (UML) 138
Unit Production Cost (UPC) 51, 171, 204, 228
Unmanned Air Vehicle (UAV) 185, 208, 224
Use Case Conversion Points (UCCP) 25, 136, 139
User Interface *see* Graphical User Interface (GUI)
update 182
upgrade 162, 181, 212
upkeep 181

V
V diagram 207
validation 37, 49, 76, 95, 109, 132, 239, 256, 293
vendors 3, 7, 51, 78, 218, 253, 273, 284

W
Warfighter Information Network 210, 218
Web based training 108, 285
Webb, Darryl 93
Weighted Methods per Class (WMC) 132
White Papers 7, 285
Whole Life Cost (WLC) 174, 183, 197, 268
Winning Productivity *see* productivity
Work Breakdown Structure (WBS) xviii, 52, 58, 87
Work package (WP) 31, 58, 87
world class 35
workflow 97, 288
Working together 97, 208
Wright, T. P. 291

X
XML 97, 288